PURE AND APPLIED MATHEMATICS

A Program of Monographs, Textbooks, and Lecture Notes

LECTURE NOTES
IN PURE AND APPLIED MATHEMATICS

1. *N. Jacobson,* Exceptional Lie Algebras
2. *L.-Å. Lindahl* and *F. Poulsen,* Thin Sets in Harmonic Analysis
3. *I. Satake,* Classification Theory of Semi-Simple Algebraic Groups
4. *F. Hirzebruch, W.D. Neumann,* and *S.S. Koh,* Differentiable Manifolds and Quadratic Forms
5. *I. Chavel,* Riemannian Symmetric Spaces of Rank One
6. *R.B. Burckel,* Characterization of C(X) among Its Subalgebras
7. *B.R. McDonald, A.R. Magid,* and *K.C. Smith,* Ring Theory: Proceedings of the Oklahoma Conference
8. *Yum-Tong Siu,* Techniques of Extension of Analytic Objects
9. *S.R. Caradus, W.E. Pfaffenberger, and Bertram Yood,* Calkin Algebras and Algebras of Operators on Banach Spaces
10. *Emilio O. Roxin, Pan-Tai Liu,* and *Robert L. Sternberg,* Differential Games and Control Theory
11. *Morris Orzech and Charles Small,* The Brauer Group of Commutative Rings
12. *S. Thomeier,* Topology and Its Applications

Other volumes in preparation

TOPOLOGY AND ITS APPLICATIONS

TOPOLOGY AND ITS APPLICATIONS

Proceedings of a Conference held at
Memorial University of Newfoundland
St. John's, Canada

Edited by

S. Thomeier

Professor of Mathematics
Memorial University of Newfoundland
St. John's, Canada

MARCEL DEKKER, INC. New York

MARCEL DEKKER, INC.

270 Madison Avenue, New York, New York 10016

LIBRARY OF CONGRESS CATALOG CARD NUMBER: 74-31691

ISBN: 0-8247-6212-6

Current printing (last digit):
10 9 8 7 6 5 4 3 2 1

PRINTED IN THE UNITED STATES OF AMERICA

PREFACE

This volume contains the proceedings of the Conference on Topology and its Applications held at Memorial University of Newfoundland from May 7 to 11, 1973. It consists of the papers given by the invited main speakers P.J. Hilton, E. Klein, A. Liulevicius and R. Thom, and of some of the shorter contributed papers presented by other participants. In those cases where a contributed paper has not been included in these proceedings, its abstract is included instead. The manuscript of René Thom's three one-hour lectures on Catastrophe Theory was prepared from audio tapes of his lectures, and the editorial work there was limited to making absolutely necessary changes only in order to preserve the flavour of the oral presentation; for valuable assistance in this task I want to thank Richard L.W. Brown as well as René Thom himself.

As the organizer and chairman of the Conference, I wish to express my appreciation to the National Research Council of Canada, to A.P.I.C.S. (the Atlantic Provinces Inter-University Committee on the Sciences) and to Memorial University for financial support of the Conference, and I want to thank all those, faculty members, graduate students and members of the secretarial staff, who gave manifold assistance before and during the Conference. My special thanks go to the Invited Speakers and the many participants whose contributions made the Conference a success.

As the editor of these proceedings, I want to express my thanks to all colleagues who assisted in the editorial tasks and the refereeing, in particular to Peter Hilton and Arunas Liulevicius. Finally, I want to thank Mrs. H. Tiller for typing a major part of the final typed copy for offset printing.

S. Thomeier

St. John's
February 1974

TABLE OF CONTENTS

TOPOLOGY AND ITS APPLICATIONS

CHARACTERISTIC NUMBERS

Arunas Liulevicius

The University of Chicago

The lectures are organized as follows: in the first lecture the algebra of unoriented cobordism is described, the second lecture applied cobordism techniques to the geometric question of immersing closed manifolds into Euclidean space, and the third introduces equivariant Stiefel-Whitney classes and numbers and applies these notions to the study of equivariant immersions of G-manifolds into representations of G. Lecture 1 is intended to popularize the work of Thom [19], Newton and VandeVelde [20], lecture 2 -- the work of Brown [4] and Liulevicius [16], lecture 3 - of tom Dieck [8], Stong [18] and Bix [2]. Only lectures 2 and 3 were presented at the conference since fog delayed my arrival.

I am grateful to the organizers of the conference for the opportunity to participate. Thanks also go to Air Canada for a fascinating tour: it was an offer which I couldn't refuse.

Lecture 1. The algebra of cobordism

We shall prove a theorem about Hopf algebras due to Newton (who worked in the context of symmetric polynomials). We prove it for a general ring A (and use the notation of Chern classes) and then specialize it to $A = Z_2$ and show how it applies to cobordism and characteristic numbers. Finally we describe the homotopy part of results of Thom [19] on the structure of the unoriented cobordism ring.

Let A be a commutative ring with unit, C a graded Hopf algebra over A which is described as follows: as an algebra $C = A[c_1,\dots,c_n,\dots]$ is a polynomial algebra on an infinite family of generators $c_n \in C^{dn}$

where $d = 1$ if the characteristic of A is 2, $d = 2$ if char $A \neq 2$. The diagonal map $\psi: C \to C \otimes C$ is given by $\psi(c_n) = \sum_{i+j=n} c_i \otimes c_j$, where of course $c_0 = 1$, the augmentation $\varepsilon : C \to A$ is described by letting it be a homomorphism of algebras over A and setting $\varepsilon(c_0) = 1$, $\varepsilon(c_n) = 0$ if $n \geq 1$; the unit map $\eta: A \to C$ is the map of algebras specified by $\eta(1) = 1 = c_0$. Define the graded dual C_* of C by setting $C_{*k} = \mathrm{Hom}_A(C^k, A)$, then $(C_*, \psi_*, \phi_*, \eta_*, \varepsilon_*)$ is a Hopf algebra. If $E = (e_1, \ldots, e_r, \ldots)$ is a sequence of natural numbers such that all but a finite number of the e_r are zero, let $c^E = c_1^{e_1} \ldots c_r^{e_r} \ldots$ and define $y_n \in C_{*dn}$ by the condition $\langle y_n, c^E \rangle = 1$ if $E = (n,0,0,\ldots)$, that is $c^E = c_1^n$, and $\langle y_n, c^E \rangle = 0$ if $E \neq (n,0,0,\ldots)$.

Theorem 1 (I. Newton). The algebra map $f: C \to C_*$ defined by $f(c_n) = y_n$ is an isomorphism of Hopf algebras.

The main application of this result is to homology of classifying spaces. Let $U = \lim_n U(n)$ be the infinite unitary group, BU its classifying space, then $H^*(BU;Z) = Z[c_1, \ldots, c_n, \ldots]$ as a polynomial algebra where $c_n \in H^{2n}(BU;Z)$ are the Chern classes. Whitney sum of vector bundles gives a map $\psi: BU \times BU \to BU$ which makes BU into an associative H-space, $\psi^* c_n = \sum_{i+j=n} c_i \otimes c_j$, so in this notation $C = H^*(BU;Z)$ for $A = Z$, $d = 2$. We have:

Corollary 2. The homology $H_*(BU;Z)$ is a polynomial algebra on classes $y_n \in H_{2n}(BU;Z)$ coming from $H_{2n}(CP^\infty;Z)$, moreover the coproduct ϕ_* is given by $\phi_*(y_n) = \sum_{i+j=n} y_i \otimes y_j$.

Proof. Only the remark about y_n coming from the homology of CP^∞ needs explanation. The standard inclusion $U(1) \xrightarrow{i} U$ induces a map $CP^\infty = BU(1) \xrightarrow{Bi} BU$, and $(Bi)^* c_n = 0$ if $n \neq 0,1$, $(Bi)^* c_1 = y$, the fundamental class of CP^∞. Let $y_n \in H_{2n}(CP^\infty;Z)$ be the class in homology dual to y^n, then under the monomorphism $(Bi)_*$ this y_n corresponds to $y_n \in C_{*2n}$ given by Theorem 1.

Motivated by this example we can ask about the filtration of $H_*(BU;Z)$ by means of the images of $H_*(BU(n);Z)$ under the standard maps $BU(n) \to BU$. This corresponds in the setting of Theorem 1 to the following: let $\Delta_i: C_* \to C_{*-di}$ be the map dual to multiplication by c_i. Define F_nC_* by: $F_nC_* = \{x \in C_* \mid \Delta_i(x) = 0 \text{ for } i > n\}$. The structure of

F_nC_* is given by:

Theorem 3. The subgroup F_nC_* is a free A-module with basis y^E $E = (e_1,\ldots,e_r,\ldots)$, $e_1 + \ldots + e_r + \ldots \leq n$.

Our first order of business is the proof of Theorem 1. We do this in a sequence of lemmas.

Lemma 4. If the element $y_n \in C_{*dn}$ is defined by $\langle y_n, c^E\rangle = 1$ if $E = (n,0,0,\ldots)$, $= 0$ if $E \neq (n,0,0,\ldots)$, then

$$\phi_*(y_n) = \sum_{i+j=n} y_i \otimes y_j .$$

Proof. We have $\langle \phi_*(y_n),\ c^{E_1} \otimes c^{E_2}\rangle = \langle y_n, c^{E_1+E_2}\rangle = 0$ unless $E_1 = (i,0,0,\ldots)$, $E_2 = (n-i,0,0,\ldots)$ in which case the value is 1.

Remark. Lemma 4 says that the algebra map $f:C \to C_*$ defined by $f(c_n) = y_n$ is a map of Hopf algebras.

Let $P(C)$ be the primitives of C: these are elements $x \in C^k$, $k > 0$ such that $\psi(x) = x \otimes 1 + 1 \otimes x$.

Lemma 5. $P(C)^{dn}$ is A-free and a direct summand of C^{dn} on one generator s_n defined recursively by

$$s_n - c_1s_{n-1} + c_2s_{n-2} - \ldots + (-1)^{n-1}c_{n-1}s_1 + (-1)^n nc_n = 0.$$

Proof. Consider the homomorphism of graded algebras $h:C \to A[x_1,\ldots,x_n]$ where x_i have grade d, defined by $h(c_k) = \sigma_k$, the k-th elementary symmetric function of $x_1,\ldots,x_n$. Of course, $h(c_k) = 0$ if $k > n$, but h is a monomorphism in gradings less than or equal to dn, and the image of h is precisely the subalgebra of all symmetric polynomials in $x_1,\ldots,x_n$. An A-free basis for the image of h is given by the elements $X(\omega)$, where $\omega = (e_1,\ldots,e_s)$ is a partition, $n \geq s \geq 0$, $e_1 > \ldots > e_s > 0$. and $X(\omega)$ is the sum of the distinct monomials obtained from $x_1^{e_1}\ldots x_s^{e_s}$ by applying permutations of $1,\ldots,n$ to the subscripts. Under the homomorphism h the diagonal $\psi:C \to C \otimes C$ corresponds to the map $\psi X(\omega) = \sum_{(\omega',\omega'')=\omega} X(\omega') \otimes X(\omega'')$, the sum ranging over all subpartitions of ω. In particular, the primitive elements under h correspond to the direct summand generated by $X((k))$, since for $k > 0$ the only partition of k having only trivial subpartitions is (k). By definition $X((k)) = x_1^k + \ldots + x_n^k$. Let

$p(x) = (x-x_1)\ldots(x-x_n) = x^n - \sigma_1 x^{n-1} + \sigma_2 x^{n-2} + \ldots + (-1)^n \sigma_n$, then $p(x_i) = 0$, $i = 1,\ldots,n$ and so

$$\begin{aligned} 0 &= p(x_1) + \ldots + p(x_n) \\ &= X((n)) - \sigma_1 X((n-1)) + \ldots + (-1)^{n-1}\sigma_{n-1}X((1)) + (-1)^n n\sigma_n. \end{aligned}$$

Let $s_k \varepsilon P(C)^{dk}$ be defined by $h(s_k) = X((k))$, then $0 = s_n - c_1 s_{n-1} + \ldots + (-1)^{n-1} c_{n-1} s_1 + (-1)^n n c_n$, and the lemma is proved.

Corollary 6. For each n, $\langle y_n, s_n \rangle = 1$.

Proof. True for $n = 1$, and if true for $n-1$ then the recursion relation for s_n shows that $s_n = c_1^n + \sum a_E c^E$, $E \neq (n,0,0,\ldots)$.

Lemma 7. For each n

$$\langle y^E, c_n \rangle = \begin{cases} 1 & \text{if } E = (n,0,0,\ldots), \\ 0 & \text{otherwise.} \end{cases}$$

Proof. True for $n = 1$. We have $\langle y_1^n, c_n \rangle = \langle y_1^{n-1} \otimes y_1, \psi(c_n) \rangle = \langle y_1^{n-1}, c_{n-1} \rangle \langle y_1, c_1 \rangle = 1$, and if $E \neq (n,0,0,\ldots)$, $y^E = y_i z$, $i > 1$, but then

$$\langle y^E, c_n \rangle = \langle y_i \otimes z, c_i \otimes c_{n-i} \rangle = 0,$$

since $\langle y_i, c_i \rangle = 0$ for $i > 1$.

Corollary 8. For each n, $\langle f(s_n), c_n \rangle = 1$.

Proof. See Corollary 6 and Lemma 7.

We are now ready to prove Theorem 1. First, f is a monomorphism. We do this by induction on the grading. Since $f(1) = 1$, f is a monomorphism in grading 0. Suppose $\operatorname{Ker} f|C_i = 0$ for $i \leq n$ and suppose $x \varepsilon \operatorname{Ker} f|C_{n+1}$. Since f is a map of Hopf algebras, $\psi(x) - x \otimes 1 - 1 \otimes x \varepsilon \operatorname{Ker} f \otimes f$, where

$$f \otimes f\colon \bigoplus_{i=1}^{n} C^i \otimes C^{n+1-i} \to C_* \otimes C_*$$

is a monomorphism, since C_* is A-free, so $\psi(x) = x \otimes 1 + 1 \otimes x$, or $x = \lambda s_k$ $(dk = n+1)$, but $\lambda = \lambda\langle f(s_k), c_k \rangle = \langle f(\lambda s_k), c_k \rangle = \langle f(x), c_k \rangle = 0$, so $x = 0$ and $f|C_{n+1}$ is a monomorphism.

To prove that f is an epimorphism, it is sufficient that $f: Q(C) \to Q(C_*)$, where $Q(B) = \overline{B}/\overline{B}\cdot\overline{B}$ denotes the indecomposable elements of a graded connected algebra B (here $\overline{B}$ = all $x \in B$ with grade $x > 0$). Since $P(C)$ is a direct summand of C, the exactness of

$$0 \longrightarrow P(C) \longrightarrow \overline{C} \xrightarrow{\overline{\psi}} \overline{C} \times \overline{C}$$

implies the exactness of

$$\overline{C}_* \otimes \overline{C}_* \xrightarrow{\overline{\psi}_*} \overline{C}_* \longrightarrow P(C)_* \longrightarrow 0 \, ,$$

so $P(C)_* = Q(C_*)$, and $\langle f(c_n), s_n \rangle = \langle y_n, s_n \rangle = 1$ by Corollary 6, so $f: C \to C_*$ is onto as well, completing the proof of Theorem 1.

Now to prove Theorem 3. Let $\Delta_i: C_* \to C_{* - di}$ be the map dual to multiplication by c_i. We have:

Lemma 9. The maps Δ_i are group homomorphisms and satisfy

$$\Delta_k(xy) = \sum_{i+j=k} \Delta_i(x)\Delta_j(y).$$

Proof. The coproduct of c_k is $\sum_{i+j=k} c_i \otimes c_j$.

We introduce a polynomial variable s and define $\Delta: C_* \to C_*[s]$ by setting $\Delta(x) = \sum_{i=0} (\Delta_i x) s^i$, then if we give the variable s the grading d (remember, $d = 1$ if char $\Lambda = 2$, $d = 2$ if char $\Lambda \neq 2$) we have:

Corollary 10. The map $\Delta: C_* \to C_*[s]$ is a homomorphism of graded algebras.

Since we now know that $C_* = \Lambda[y_1, \ldots, y_n, \ldots]$ we can introduce an additional bit of structure into C_*. If $E = (e_1, \ldots, e_r)$ is a sequence of natural numbers, let (as before) $y^E = y_1^{e_1} \cdots y_r^{e_r}$ and define $\deg y^E = e_1 + \ldots + e_r$ and set $\deg(\sum a_E y^E)$ = maximum $\deg y^E$ where $a_E \neq 0$. The invariant deg is the algebraic degree in the polynomial generators y_i and is to be distinguished from the grading: recall that grade $y^E = d(e_1 + 2e_2 + \ldots + re_r)$. We also have the notion of degree in $C_*[s]$, namely $\deg(\sum a_i s^i)$ = maximum i such that $a_i \neq 0$.

Lemma 11. The map $\Delta: C_* \to C_*[s]$ preserves degree.

Proof. We have $\Delta y_k = y_k + y_{k-1}s$, so if $E = (e_1, \ldots, e_r)$, $y^E = y_1^{e_1} \cdots y_r^{e_r}$, then $\Delta y^E = (\Delta y_1)^{e_1} \cdots (\Delta y_r)^{e_r}$ has degree $e_1 + \ldots + e_r$

in s, and the top coefficient is $y_1^{e_2}\cdots y_{r-1}^{e_r}$, so two monomials y^E, $y^{E'}$ of the same grade have the same top coefficient in $\Delta(y^E)$ and $\Delta(y^{E'})$ if and only if $(e_2,\ldots,e_r) = (e'_2,\ldots,e'_r)$, but this implies $e_1 = e'_1$ as well, so $y^E = y^{E'}$.

We define a filtration on C_* by setting $F_nC_* = \{x \in C_* | \deg \Delta x \leq n\}$, that is $x \in F_nC_*$ if and only if $\Delta_i = 0$ for $i > n$, we have:

Theorem 3. F_nC is the free A-module on monomials y^E where $\deg y^E \leq n$.

Examples. 1. The standard map $BU(n) \to BU$ induces a monomorphism $H_*(BU(n);Z) \to H_*(BU;Z)$ and the image is precisely the free abelian group on y^E with $\deg y^E \leq n$.

2. The standard map $BO(n) \to BO$ induces a monomorphism $H_*(BO(n);Z_2) \to H_*(BO;Z_2)$ and the image is precisely the subspace over Z_2 with basis x^E, $\deg x^E \leq n$, where x_i is the element in $H_i(BO(1);Z_2)$ dual to w_1^i.

3. The operations Δ_i give us a quick way of determining the incidence matrices of C with C_* (VandeVelde [20] has them explicitly for $H_n(BU;Z)$ for $n \leq 24$). For example, $\langle c_2, y_1^2\rangle = 1$, $\langle c_1^2, y_1^2\rangle = \langle c_1, \Delta_1(y_1^2)\rangle = \langle c_1, 2y_1\rangle = 2$ and we have the incidence matrix

	c_2	c_1^2
y_2	0	1
y_1^2	1	2

Indeed, VandeVelde [20] uses the triangularity of the incidence matrices under a clever ordering of the c^E, y^F bases to give a different proof of Theorem 1.

We now explain how this bears on the algebra of unoriented cobordism (see the original paper of Thom [19]). Let $\alpha: E(\alpha) \to X$ be a vector bundle with structure group $O(n)$ and $EB(\alpha)$ the total space of the unit ball bundle, $ES(\alpha)$ the total space of the unit sphere bundle, $M(\alpha) = EB(\alpha)/ES(\alpha)$ the Thom space of α. Notice that $M(\alpha \times \beta) = M(\alpha) \wedge M(\beta)$, $M(\varepsilon^n) = S^n$, where $\varepsilon^n: R^n \to$ point. The reduced Z_2-cohomology of $M(\alpha)$ is a free $H^*(X;Z_2)$-module on one generator

$U \in \tilde{H}^n(M(\alpha);Z_2)$ (dim $\alpha = n$) and the map

$$\phi : H^k(X;Z_2) \to \tilde{H}^{n+k}(M(\alpha);Z_2)$$

defined by $\phi(x) = x \cdot U$ is an isomorphism (U is called the Thom class of α and ϕ is called the Thom isomorphism). Of course, dually we have the Thom isomorphism in homology:

$$\phi_* : \tilde{H}_{n+k}(M(\alpha);Z_2) \to H_k(X;Z_2).$$

We write $MO(n) = M(\gamma^n)$, where γ^n is the classifying n-plane bundle over $BO(n)$, and we have maps

$$e_n : MO(n) \wedge S^1 \to MO(n+1)$$

$$\mu_{m,n} : MO(m) \wedge MO(n) \to MO(m+n)$$

induced by the standard inclusion $O(n) \to O(n+1)$ and the Whitney sum representation $O(m) \times O(n) \xrightarrow{w} O(m+n)$. Using the suspension homomorphisms and maps induced by e_n we define

$$\pi_m(MO) = \lim_{\substack{\to \\ n}} \pi_{m+n}(MO(n)) \quad ,$$

$$H_m(MO;Z_2) = \lim_{\substack{\to \\ n}} \tilde{H}_{m+n}(MO(n);Z_2).$$

Since the following diagram commutes

$$\begin{array}{ccc} \tilde{H}_{m+r}(MO(m);Z_2) \otimes \tilde{H}_{n+s}(MO(n);Z_2) & \xrightarrow{\mu_{m,n*}} & H_{m+n+r+s}(MO(m+n);Z_2) \\ \Big\downarrow \phi_* \otimes \phi_* & & \Big\downarrow \phi_* \\ H_r(BO(m);Z_2) \otimes H_s(BO(n);Z_2) & \xrightarrow{w_*} & H_{r+s}(BO(m+n);Z_2) \end{array}$$

(here ϕ_* is the Thom isomorphism in homology), the maps $\mu_{m,n}$ induce a ring structure in $\pi_*(MO)$, $H_*(MO;Z_2)$ and $\phi_* : H_*(MO;Z_2) \to H_*(BO;Z_2)$ is a ring isomorphism. We define $b_n \in H_n(MO;Z_2)$ by $\phi_*(b_n) = x_n$. Notice that b_n is born on $MO(1) \equiv BO(1)$ and there has the name x_{n+1}, the class dual to w_1^{n+1}. The Hurewicz homomorphisms over Z_2 $h:\pi_{m+k}(MO(k)) \to H_{m+k}(MO(k);Z_2)$ fit together to give a ring homomorphism $h:\pi_*(MO) \to H_*(MO;Z_2)$. Let A_* be the dual of the Steenrod algebra over Z_2 (see Milnor [17]), $\mu:H_*(MO;Z_2) \to A_* \otimes H_*(MO;Z_2)$

the coaction. Since μ is a homomorphism of algebras over Z_2 it is sufficient to specify $\mu(b_n)$. Now

$$\mu_*(b_n) = \sum_{s=0}^{s} \gamma_{n-s}^{(s+1)} \otimes b_s$$

where $\gamma_{n-s}^{(s+1)} \varepsilon A_{*n-s}$ satisfy the relations:

$$\gamma_s^{(1)} = \begin{cases} \xi_r & \text{if } s = 2^r - 1 , \\ 0 & \text{if } s \neq 2^r - 1 , \end{cases}$$

where ξ_r are the Milnor [17] generators, moreover $\gamma_0^{(n+1)} = 1$ and the Cartan relations are satisfied: for each pair of natural numbers i,j we have

$$\gamma_r^{(i+j)} = \sum_{r=s+t} \gamma_s^{(i)} \gamma_t^{(j)}$$

(see Liulevicius [15], for example).

Theorem 12 (Thom). The algebra $\pi_*(MO)$ is a polynomial algebra $Z_2[u_2,u_4,\ldots,u_n,\ldots]$, $n \neq 2^r - 1$ and $h:\pi_*(MO) \to H_*(MO;Z_2)$ is a monomorphism onto the elements x such that $\mu(x) = 1 \otimes x$ (the primitives under the coaction μ).

Proof. Let $N_* = Z_2[u_2,u_4,\ldots,u_n,\ldots]$, $n \neq 2^r - 1$ and define a homomorphism of algebras and comodules over A_*

$$f:H_*(MO;Z_2) \to A_* \otimes N_*$$

(the target being the extended A_*-comodule on N_*) by setting $\underline{f} = (\eta_* \otimes 1)f:H_*(MO;Z_2) \to N_*$ $\underline{f}(b_n) = u_n$ if $n \neq 2^r - 1$, $\underline{f}(b_n) = 0$ if $n = 2^r - 1$ for some r. Then since $f = (1 \otimes \underline{f})\mu$ we have $f(b_n) = 1 \otimes u_n$ modulo decomposables if $n \neq 2^r - 1$, $f(b_{2^r-1}) = \xi_r \otimes 1$ modulo decomposables, so f is onto, hence an isomorphism, since the dimensions of the domain and target are the same in each grading. Since $H_*(MO;Z_p) = 0$ for p an odd prime it follows that MO is equivalent to the Eilenberg-MacLane spectrum on the graded Z_2 vector space N_*, and the theorem follows.

Remark. This version of the proof appears in Liulevicius [14] (see also the correction in [15]). Of course, the image of the Hurewicz homomorphism is precisely $f^{-1}(1 \otimes N_*)$.

Let us simplify notation by identifying $H_*(MO;Z_2)$ with $A_* \otimes N_*$ under f, thus identifying $\pi_*(MO)$ with $1 \otimes N_*$. The following table gives the u_n in terms of b_i for $n \leq 10$ (for u_n, $n \leq 18$ see Table 1.1 in Liulevicius [16]).

Table

generator	algebraic degree	expression
ξ_1	1	b_1
u_2	1	b_2
ξ_2	1	b_3
	2	$+ b_1b_2$
u_4	1	b_4
	3	$+ b_1^2b_2$
u_5	1	b_5
	2	$+ b_1b_4 + b_2b_3$
	3	$+ b_1b_2^2$
u_6	1	b_6
ξ_3	1	b_7
	2	$+ b_3b_4 + b_1b_6$
	3	$+ b_1b_2b_4 + b_1^2b_5$
	4	$+ b_1^2b_2b_3 + b_1^3b_4$
	5	$+ b_1^3b_2^2$
u_8	1	b_8
	3	$+ b_2b_3^2 + b_1^2b_6$
u_8	1	b_8
	3	$+ b_2b_3^2 + b_1^2b_6$
	5	$+ b_1^2b_2^3 + b_1^4b_4$
	7	$+ b_1^6b_2$

Lecture 2. Immersions up to cobordism

There is a useful rule of thumb in differential topology for studying complicated geometric structures. Usually there is a wealth of structure so to understand it one tries to simplify the situation by deciding which parts of structure can be pruned off. The aim is to reduce the tangled, continuous picture of geometrical reality to the tangled but discrete domain of homotopy. Many geometers consider the problem solved at this point, by definition: if homotopy information is required, one just picks up the phone and calls Mahowald. Suppose, however, that you are stranded on a desert island -- you have to untangle the homotopy yourself. The technique of course is to reduce it to a problem of algebra. If the algebraic problem is still too complicated to handle it may be necessary to look if certain aspects of the homotopy situation can be simplified so that the algebra becomes manageable.

The work of Thom [19] on unoriented cobordism may be taken as an example (at the risk of making the scheme into a Procrustean bed, of course.) Suppose we want to get an overall view of the class of all closed smooth manifolds. The first problem is that there are too many of them even under diffeomorphism: there are lots and lots of those which are even topologically spheres, so the relation of diffeomorphism is too strong -- there are too many equivalence classes. We need a weaker, but hopefully still interesting, equivalence relation to cut down the number of equivalence classes. Cobordism, born in Poincaré's first attempt to define homology, is a reasonable candidate, especially if we are interested in homology of manifolds. Two closed manifolds M and M' of dimension m are said to be cobordant if there exists an $(m+1)$-dimensional manifold with boundary W such that ∂W is diffeomorphic to the disjoint sum of M with M'. It is immediately checked that cobordism is an equivalence relation (use the collaring theorem for transitivity). We denote the set of equivalence classes of m-dimensional closed manifolds under cobordism by N_m. Disjoint sum induces addition in N_m which makes it into a vector space over Z_2, and Cartesian product induces a product $N_m \otimes N_n \to N_{m+n}$ which makes $N_* = \{N_m\}_{m \in Z}$

into a graded algebra. The problem is now to determine the structure of this algebra, and the key to Thom's solution is a reduction of the geometric problem to a problem in homotopy, namely $\pi_*(MO)$. We exhibit the Thom-Pontrjagin map $\tau: N_m \to \pi_m(MO)$. Let M be an m-dimensional manifold. Choose an embedding $e:M^m \to R^{m+k}$. Let T be a tubular neighborhood of the embedding $e : T$ is diffeomorphic to $EB(\nu)$, the total space of the unit ball bundle associated with the normal bundle ν of the embedding e; let $d:T \to EB(\nu)$ be the diffeomorphism and notice that under d, ∂T corresponds to $ES(\nu)$, the total space of the unit sphere bundle associated with ν. Consider $S^{m+k} = R^{m+k} \cup \{\infty\}$ as the one-point compactification of R^{m+k}, let $t:S^{m+k} \to T/\partial T$ be the map induced by the identity on T which maps $S^{m+k} - T$ into the point of $T/\partial T$ represented by ∂T, let $\overline{d} : T/\partial T \to EB(\nu)/ES(\nu) = M(\nu)$ be the map induced by d, and $\hat{\nu}: M(\nu) \to M(\gamma^k) = MO(k)$ be the map induced by a classifying map of ν. If $x \in N_m$ is the cobordism class defined by M, let $\tau(x) \in \pi_m(MO)$ be the class defined by the composition

$$S^{m+k} \xrightarrow{t} T/\partial T \xrightarrow{\overline{d}} M(\nu) \xrightarrow{\hat{\nu}} MO(k).$$

Of course, one has to verify that $\tau(x)$ is independent of all the choices made in its construction, but this is not difficult. It is almost immediate that τ is an algebra map, and one shows that τ is an isomorphism by explicitly constructing an inverse $\pi_m(MO) \to N_m$ (here the transversality theorems of Thom are used).

This reduces the geometry to homotopy, and we consider the problem solved, since the algebra structure of $\pi_*(MO)$ has been determined: it is a polynomial algebra $Z_2[u_2,u_4,u_5,\ldots,u_n,\ldots]$, $n \neq 2^r - 1$. We would like to identify the monomorphism

$$N_m \xrightarrow{\tau} \pi_m(MO) \xrightarrow{h} H_m(MO;Z_2) \xrightarrow{\phi_*} H_m(BO;Z_2)$$

and this is easily done: it is the <u>normal characteristic number map</u> $\nu_\#$ which is described as follows: let $u \in H^m(BO;Z_2)$, so we may consider u as a linear functional on $H_m(BO;Z_2)$, then

$$u\phi_*h\tau(\text{class of } M) = \langle\nu^*(u),[M]\rangle ,$$

where $\nu:M \to BO$ is the stable normal bundle of M, $[M]$ is the fundamental class of M, and $\langle,\rangle$ is the Kronecker pairing between cohomology and homology. For example, if $M = RP^2$ then $W(\nu) = 1 + x$, so

$<w_2(\nu),[RP^2]> = 0$, $<w_1^2(\nu),[RP^2]> = 1$, and $\phi_* h\tau(\text{class of } RP^2) = x_2$, the second standard polynomial generator of $H_*(BO;Z_2)$.

Let us now come to our geometric problem and see whether it is possible to say something reasonable about it using the procedure sketched above. Given a closed manifold M^m we would like to know the least k such that M^m immerses into R^{m+k}. According to Whitney, $k \leq m-1$ if $m \geq 2$. The immersion problem has been studied intensively for special classes of manifolds and a great deal is known (see Gitler [11]). There is a conjecture that (if $m \geq 2$) M^m immerses in $R^{2m-\alpha(m)}$, where $\alpha(m)$ is the number of ones in the dyadic expansion of m.

We shall say that <u>M^m immerses into R^{m+k} up to cobordism</u> if there is a manifold M' cobordiant to M such that M' immerses in R^{m+k}. Brown [4],[5] has proved that M^m immerses in $R^{2m-\alpha(m)}$ up to cobordism. His technique is to exhibit polynomial generators for the unoriented cobordism ring N_* which satisfy this condition. Given M^m, let $k(M)$ be the least integer such that M^m immerses up to cobordism in $R^{m+k(M)}$. Of course, we have lost a lot of geometric information by passing to cobordism, since now $k(M)$ is a function only of the cobordism class of M, and if M is a boundary (for example $M = RP^{2n+1}$, an odd-dimensional real projective space) then $k(M) = 0$. Even if M is not a boundary, a manifold may immerse up to cobordism into a lower dimensional Euclidean space than M itself; for example, RP^{10} immerses up to cobordism into R^{15}, but RP^{10} itself immerses into R^{16}, and does not immerse into R^{15} (see Gitler [11]).

We should explain how we have reduced the geometric problem to a question of homotopy. Let $MO = \{MO(n),e_n\}$ be the Thom spectrum for the orthogonal group (see Lecture 1). Let $\pi^{st}_{m+k}(MO(k))$ be the stable $(m+k)$-th homotopy group of $MO(k)$ and

$$\lambda_k : \pi^{st}_{m+k}(MO(k)) \to \pi_m(MO)$$

the map into the direct limit.

<u>Theorem 1</u>. If $x \in \pi_m(MO)$ represents the cobordism class of M^m, then M immerses up to cobordism into R^{m+k} if and only if x is in the image of λ_k.

<u>Proof</u>. Suppose M is cobordant to M' which immerses in R^{m+k} with normal bundle ν, then there is an embedding $e:M' \to R^{m+k+s}$ with

normal bundle $\nu \oplus s$ (where s denotes the trivial s-dimensional bundle over M') with s sufficiently large, and the Thom map τ(class M) is represented by (see Lecture 1 for notation)

$$S^{m+k+s} \xrightarrow{t} T/\partial T \xrightarrow{\overline{d}} M(\nu) \wedge S^s \xrightarrow{\hat{\nu}\wedge 1} MO(k) \wedge S^s \longrightarrow MO(k+s)$$

so τ(class M) is in the image of λ_k.

Conversely, suppose $x \in$ image λ_k, so for s sufficiently large, x is represented by $f:S^{m+k+s} \to MO(k) \wedge S^s \to MO(k+s)$, so by the usual Thom transversality argument f is homotopic to a map g such that $g^{-1}(BO(k+s))$ is a closed submanifold M of S^{m+k+s} with normal bundle $\nu \oplus s$, where ν is a k-plane bundle over M. By a theorem of Hirsch [12] M immerses in R^{m+k} and of course τ(class M) = x.

Stated in another way: we define an increasing filtration (called the geometric filtration) of $\pi_*(MO)$ by setting ${}^{geo}F_k\pi_*(MO)$ = image λ_k. Let ${}^{geo}E^0_*\pi_*(MO)$ be the associated graded object, then our original question is equivalent to the following: given x = (class $(M^m) \neq 0$, what is the k so that the class of x is non-zero in ${}^{geo}E^0_{k,m-k}\pi(MO)$? The purpose of this lecture is to study ${}^{geo}E^0_{*,*}\pi(MO)$. The fact that we have lost geometric information by passing to cobordism (and reducing the problem to homotopy theory) should not make us sad: the geometric situation was too complicated, so we would not obtain useful qualitative information if we had preserved the complexity of the original geometric problem. We will soon see that there is still a great deal (possibly too much) structure remaining.

Let us sum up what we have learned about the unoriented cobordism ring $N_* = \pi_*(MO)$ in Lecture 1: $\pi_*(MO) = Z_2[u_2,u_4,\ldots,u_n,\ldots]$, $n \neq 2^r - 1$ on a set of "algebraically obvious" generators u_n, the mod 2 Hurewicz homomorphism $h:\pi_*(MO) \to H_*(MO;Z_2)$ is a monomorphism onto the primitives under the coaction $\mu:H_*(MO;Z_2) \to A_* \otimes H_*(MO;Z_2)$ under the dual of the Steenrod algebra. We have the commutative diagram

$$\begin{array}{ccc} \pi^{st}_{m+k}(MO(k)) & \xrightarrow{\lambda_k} & \pi_m(MO) \\ \downarrow h & & \downarrow h \\ \tilde{H}_{m+k}(MO(k);Z_2) & \xrightarrow{\ell_k} & H_m(MO;Z_2) \end{array}$$

where ℓ_k is the direct limit map in homology. We define the <u>algebraic filtration</u> ${}^{alg}F$ of $\pi_*(MO)$ by setting ${}^{alg}F_k = h^{-1}(\text{image } \ell_k)$. Notice that ${}^{geo}F_k \subset {}^{alg}F_k$. The algebraic filtration is much easier to handle (courtesy of Theorem 3 of lecture 1: $x \in {}^{alg}F_k$ if and only if $h(x)$ is of degree less than or equal to k in the polynomial generators $b_1, b_2, \ldots$). We ask ourselves four questions:

<u>Question A</u>. Do the algebraically obvious polynomial generators have the minimal algebraic filtration?

This question tacitly hopes that ${}^{alg}E^0\pi_*(MO)$ has a very obvious structure: namely, take polynomial generators of smallest possible filtration, then monomials in these generators should project onto a basis of ${}^{alg}E^0\pi_*(MO)$. Unfortunately the answer to Question A turns out to be <u>no</u> (the first generator not having minimal algebraic filtration occurs in dimension 11), so we ask:

<u>Question B</u>. Do the polynomial generators of Boardman [3], Brown [4], Dold [9] and Kozma [13] have minimal algebraic filtration?

The answer is <u>no</u> again: Boardman's generators first fail in dimension 11, Brown's in dimension 6, Dold's in dimension 11, Kozma's in dimension 11.

We can still ask:

<u>Question C</u>. Is ${}^{alg}E^0\ \pi_*(MO)$ a polynomial algebra?

The answer is <u>no</u>, as can be expected. The first departure from a polynomial algebra occurs in dimension 10.

So far we have been operating under the assumption that algebraic filtration is the same as geometric filtration, but this should also be examined:

<u>Question D</u>. Is the algebraic filtration of $\pi_*(MO)$ the same as the geometric filtration?

The answer is <u>no</u> in general: we shall show that the cobordism class of RP^m for $m = 2^r - 2$ has algebraic filtration 1 for all r, but for $r \geq 4$ it has geometric filtration at least 2. Through grading 10 the algebraic and geometric filtrations coincide (indeed, preliminary computations seem to say that they coincide through grading 13).

Let us give a sketch of proof of some of these results (for fuller discussion see Liulevicius [16]).

Proposition 2. $\mathrm{alg}_E{}^0\pi_*(MO)$ is not a polynomial algebra.

Proof. There is a unique polynomial generator $u_5 = b_5 + b_1b_4 + b_2b_3 + b_1b_2^2$ of filtration 3 and two choices for the generator in grade 4: $u_4 = b_4 + b_1^2b_2$ and $u_4' = u_4 + u_2^2 = b_4 + b_2^2 + b_1^2b_2$, both choices having the minimal filtration 3. Now $u_5^2 = b_5^2 + b_1^2b_4^2 + b_2^2b_3^2 + b_1^2b_2^4$ has filtration 6, so do $u_2^3u_4 = b_2^3b_4 + b_1^2b_2^4$ and $u_2^3u'_4 = b_2^3b_4 + b_2^5 + b_1^2b_2^4$, but $u_5^2 + u_2^3u_4$ has filtration 4, $u_5^2 + u_2^3u'_4$ has filtration 5, so in both cases there is a drop in filtration.

Proposition 3. The algebraically obvious polynomial generator u_{11} has filtration 8, whereas the polynomial generator v_{11} of minimal filtration has filtration 5, moreover $v_{11} = u_2u_9 + u_{11}$.

Proof. Solve for u_9 and u_{11} as indicated at the end of Lecture 1.

Remark. The next generator u_{12} has filtration 5 (off by 2), and u_{13} has filtration 10 (off by 7).

Proposition 4. If $n = 2^r - 2$, $r \geq 4$, then the geometric filtration of the cobordism class of RP^n is at least 2.

Proof. Let S be the sphere spectrum, HZ_2 the Eilenberg-MacLane spectrum for Z_2 and Y the fiber of $S \to HZ_2$ (the map representing the generator of $H^0(S;Z_2)$). The Z_2 homology of Y is just the augmentation ideal of A_* with a shift down by one in grading: $H_k(Y;Z_2) = (A_*)_{k+1}$. Let P be the suspension spectrum of RP^∞ and let $J:P \to S$ be the stable J-map. Since $J_* = 0$ in homology, it factors through $j:P \to Y$ and $j_*(x_n) \neq 0$ for all n, where $x_n \in H_n(RP^\infty;Z_2)$ is the dual of w_1^n. In particular, if $n = 2^r - 1$ then x_n is primitive under the coaction A_*, so $j_*x_n = [\xi_1^{2^r}]$, since only non-zero primitives of A_* are the elements $\xi_1^{2^s}$ (see Adams [1]). Now remember that $MO(1)$ is homotopy equivalent to RP^∞ and under this homotopy equivalence b_n corresponds to x_{n+1}, so $j_*b_{2^r-2} = [\xi_1^{2^r}]$, and the latter is not in the image of the Z_2 Hurewicz homomorphism for $r \geq 4$ by Adams [1]. It remains to show that $h(\text{class } RP^{2^r-2}) = b_{2^r-2}$, but this is immediate, since $(1 + x)^{-2^r+1} = (1 + x)^{-2^r}(1 + x) = 1 + x$, so

$\overline{w}_i(RP^{2^r-2}) = 0$ for $i \neq 0, 1$ and $\overline{w}_1 = x$, so

$\nu_{\#}(RP^{2^r-2}) = x_{2^r-2} = \phi_* b_{2^r-2}$, as claimed.

Lecture 3. Equivariant Stiefel-Whitney classes and numbers

Let G be a finite group (more generally, a compact Lie group). Let X be a G-space, then a G-vector bundle over X is a vector bundle $\alpha:E \to X$ where E is a G-space, and for each $g \in G$ the action $g:E \to E$ is a map of vector bundles, that is, α is a G-map and g acts linearly on fibers.

We wish to construct a classifying bundle for all such G-vector bundles of dimension n, say $\gamma_n:E_n \to B(G,O(n))$. We shall follow Stong ([18],p.7). Let R^∞ denote the direct sum of a countable number of copies of each real irreducible representation of G and let $s:G \times R^\infty \to R^\infty$ be the resulting linear action of G. Let $B(G,O(n))$ be the Grassmanian of real n-dimensional subspaces of R^∞. Define a G-action on $B(G,O(n))$ in the obvious way: if π is an n-plane, $g \in G$, let $g \cdot \pi = \{g \cdot x | x \in \pi\}$. Let E_n be the set of pairs (π,x) where π is an n-plane in R^∞ and $x \in \pi$ and let $\gamma_n(\tau,x) = \pi$. The G-action on E_n is induced by the G-action on $B(G,O(n))$ and R^∞. The map $\gamma_n:E_n \to B(G,O(n))$ is a G-bundle and it is a universal n-plane G-bundle. Indeed, there is a universal principal G-bundle $E(G,O(n))$ = the Stiefel manifold of n-frames in R^∞. T.tom Dieck [6] has a very general process for constructing universal principal G-bundles for the category of numerable principal G-bundles.

We wish to introduce the notion of equivariant Stiefel-Whitney classes. We do this by using a classical idea of A. Borel to transfer ourselves from the category of G-spaces to the category of topological spaces and then take singular cohomology. This has been very fruitfully exploited by T. tom Dieck [7],[8]. Let $p:EG \to BG$ be a universal principal G-bundle (in particular, BG is homeomorphic to the orbit space EG/G and the action of G is trivial on BG). If X is a G-space, then $p \times_G 1:EG \times_G X \to BG$ is a fiber bundle over BG with fiber X. The functor which associates $EG \times_G X$ to X is a covariant functor from the homotopy category of G-spaces to the homotopy category of spaces. If X is a G-space, we let $h^q(X) = H^q(EG \times_G X; Z_2)$. To define characteristic classes of n-dimensional G-bundles in this cohomology theory we have to examine $h^*(B(G,O(n)))$. The task is made easier by the following

<u>Theorem 1 (T. tom Dieck</u>. Let the classifying bundle $p_n: EO(n) \to BO(n)$ have trivial G-action and $j: BO(n) \to B(G,O(n))$ be a classifying map for p_n, then

$$1 \times_G j: EG \times_G BO(n) \to EG \times_G B(G,O(n))$$

is a homotopy equivalence.

<u>Proof</u>. The map $j: BO(n) \to B(G,O(n))$ is a homotopy equivalence if we ignore the G-action, so $1 \times_G j$ is a map of fiber bundles over BG inducing the identity on BG and the map j on the fibers, so by Theorem 6.3 of Dold [9] the map $1 \times_G j$ is a fiber homotopy equivalence.

<u>Remark</u>. Of course, $EG \times_G BO(n) = BG \times BO(n)$ since the action of G on $BO(n)$ is trivial, and we have $h^*(B(G,O(n)) = H^*(BG) \otimes H^*(BO(n))$.

We have products in the theory $h^*(\ )$, since if X_1 and X_2 are G-spaces, we can give $X_1 \times X_2$ the diagonal G-action and the G-maps $\pi_i: X_1 \times X_2 \to X_i$ for $i = 1,2$ induce a G-map $EG \times_G (X_1 \times X_2) \to (EG \times_G X_1) \times (EG \times_G X_2)$ and hence a cup product

$$h^p(X_1) \otimes h^q(X_2) \to h^{p+q}(X_1 \times X_2).$$

Notice that this gives a cup product in $h^*(X)$ and an action of $h^*(\text{point}) = H^*(BG)$ which makes $h^*(X)$ into an algebra over $h^*(\text{point})$.

<u>Corollary 2</u>. The algebra $h^*(B(G,O(n))$ is free over $h^*(\text{point})$ on generators $Z_2[w_1,\ldots,w_n]$ where w_k corresponds to $\pi_2^*(w_k)$ under

$$EG \times_G B(G,O(n)) \xleftarrow{1 \times_G j} BG \times BO(n) \xrightarrow{\pi_2} BO(n).$$

The homotopy equivalence $1 \times_G j$ indicates that if we are given a G-vector bundle $\alpha: E \to X$ of dimension n then α should give rise to a principal G-bundle over $EG \times_G X$ and a vector bundle of dimension n. It is not difficult to identify these bundles. The construction $1 \times_G j$ indicates that the principal G-bundle should depend just on X, rather than on α. It is just the quotient map $EG \times X \to EG \times_G X$.

The vector bundle over $EG \times_G X$ is of course the map $1 \times_G \alpha$. Suppose $\mathcal{U} = \{U\}$ is a trivializing cover for $p: EG \to BG$ and $\mathcal{V} = \{V\}$ is a trivializing cover for $\alpha: E \to X$, that is, there exist G-isomorphisms $h_U: U \times G \to p^{-1}(U)$ and $k_V: V \times R^n \to \alpha^{-1}(V)$ where V is a G-invariant open subset of X and the action of G on $V \times R^n$ is given by $g(x,\beta) = (g \cdot x, \gamma(x,g)\beta)$ where $\gamma: V \times G \to O(n)$ is a map such that $\gamma(x,\ ): G \to O(n)$ is a representation for each $x \in V$. We have the useful

Lemma 3. If Y is a G-space, then the map $U \times G \times Y \to U \times Y$ given by $(u,g,y) \to (u,gy)$ induces a homeomorphism of $(U \times G) \times_G Y$ with $U \times Y$.

Proof. Immediate, since $(u,y) \to [u,e,y]$ is an inverse.

Since this homeomorphism is natural in Y, we have

$$\begin{array}{ccc} (U \times G) \times_G (V \times R^n) & \approx & U \times V \times R^n \\ \downarrow {\scriptstyle 1 \times_G \pi_1} & & \downarrow {\scriptstyle 1 \times \pi_1} \\ (U \times G) \times_G V & \approx & U \times V \end{array}$$

and it follows that $1 \times_G \alpha: EG \times_G E(\alpha) \to EG \times_G X$ is an n-plane vector bundle. The same argument shows that $1 \times_G S(\alpha) = S(1 \times_G \alpha)$, $1 \times_G B(\alpha) = B(1 \times_G \alpha)$, $1 \times_G P(\alpha) = P(1 \times_G \alpha)$ where $S(\alpha)$ is the sphere bundle of α, $B(\alpha)$ the unit ball bundle of α and $P(\alpha)$ is the projective space bundle associated to α. We can reinterpret the equivariant Stiefel-Whitney classes of α as the ordinary Stiefel-Whitney classes of $1 \times_G \alpha$. We can restate Corollary 2 as follows:

Corollary 4. $H^*(EG \times_G B(G,O(n)) = H^*(BG)[w_1,\ldots,w_n]$, $w_k = w_k(1 \times_G \gamma^n)$ is the k-th Stiefel-Whitney class of the bundle $1 \times_G \gamma^n$, γ^n being the classifying n-plane G-bundle over $B(G,O(n))$.

The nice thing about equivariant Stiefel-Whitney classes is that they help distinguish even G-vector bundles over a point. For example, if we let $G = Z_2$ and let $\rho: Z_2 \to O(1)$ be the identity map, then if we let $\alpha: R \to *$ be a Z_2-bundle over a point with the action of Z_2 given by $Tr = -r$ for the generator T, the associated line bundle $1 \times_{Z_2} \alpha: S^\infty \times_{Z_2} R \to S^\infty \times_{Z_2} *$ is precisely the canonical line bundle over RP^∞, and $w_1(1 \times_{Z_2} \alpha) = x$, the generator of $H^1(RP^\infty; Z_2)$.

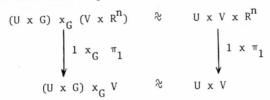

Here is another bonus of the new interpretation of equivariant Stiefel-Whitney classes: if we let $w_k^G(\alpha) = w_k(1 \times_G \alpha)$ then since $1 \times_G (\alpha + \beta) = (1 \times_G \alpha) + (1 \times_G \beta)$ and $1 \times_G (\alpha \times \beta) = (1 \times_G \alpha) \times (1 \times_G \beta)$, it follows that for example:

Corollary 5. If α, β are G-vector bundles, then

$$w_k^G(\alpha \oplus \beta) = \sum_{i+j=k} w_i^G(\alpha) \cup w_j^G(\beta).$$

Let us illustrate the notions so far introduced by looking at linear G-actions on projective spaces. The work described here is part of Michael Bix's thesis (U.C., 1974). If $\rho: G \to O(n)$ is a representation and $V = R^n$ with this action of G, then $EG \times_G V \xrightarrow{\gamma} BG$ is a vector bundle with classifying map $B\rho: BG \to BO(n)$. We shall let $w_i^G(V) = w_i(\gamma)$. Since the center of $O(n)$ is $Z_2 = \{I, -I\}$, the action of G on V induces a G-action on the projective space $P(V)$ and $EG \times_G P(V) \cong P(EG \times_G V)$. This allows us to compute $h^*(P(V))$:

Theorem 6. If $\rho: G \to O(n)$ is a representation and V is the underlying representation module then $h^*(P(V)) = H^*(EG \times_G P(V); Z_2)$ is a free $H^*(BG; Z_2)$-module on $1, a, \ldots, a^{n-1}$, where $a \in h^1(P(V))$ is $w_1(\lambda)$, $\lambda: EG \times_G S(V) \to EG \times_G P(V)$ the canonical double cover, moreover

$$a^n = w_n^G(V) \cdot 1 + w_{n-1}^G(V) a + \ldots + w_1^G(V) a^{n-1} .$$

Proof. Since $EG \times_G P(V) = P(EG \times_G V)$ we have the commutative diagram (γ^n is the universal n-plane bundle)

$$\begin{array}{ccc} P(EG \times_G V) & \longrightarrow & P(\gamma^n) \\ \downarrow & & \downarrow \\ BG & \xrightarrow{B\rho} & BO(n) \end{array}$$

and since the corresponding result is valid in $P(\gamma^n)$, the theorem follows. For the sake of completeness we also recall how the result for $P(\gamma^n)$ is proved. Recall that $P(\gamma^n) = BO(n-1) \times BO(1)$ and the class $a \in H^1(P(\gamma^n); Z_2)$ classifying the double cover $S(\gamma^n) \to P(\gamma^n)$ is given by $a = 1 \otimes w_1$. Since the fiber of $P(\gamma^n) \to BO(n)$ is RP^{n-1} and $i^*a = x$, the fundamental class of RP^{n-1}, it follows by the Leray-Hirsch theorem that $1, a, \ldots, a^{n-1}$ is a free $H^*(BO(n))$-basis for $H^*(P(\gamma^n))$. We have

$$\begin{aligned} w_n \cdot 1 + w_{n-1} \cdot a + \dots w_1 \cdot a^{n-1} \\ = w_{n-1} \otimes w_1 \\ + w_{n-1} \otimes w_1 + w_{n-2} \otimes w_1^2 \\ \vdots \\ w_2 \otimes w_1^{n-2} + w_1 \otimes w_1^{n-1} \\ + w_1 \otimes w_n^{n-1} + 1 \otimes w_1^n \\ = 1 \otimes w_1^n = a^n, \end{aligned}$$

so the equation is satisfied in $H^*(P(\gamma^n))$, and the theorem follows.

Examples. 1. Let $\rho: Z_2 \to O(n+1)$ be the standard inclusion of $O(1)$ into $O(n+1)$, that is

$$T = \begin{bmatrix} -1 & & & \\ & 1 & & \\ & & \ddots & \\ & & & 1 \end{bmatrix}$$

and denote the induced action on RP^n by σ_1, then

$$h^*(RP^n, \sigma_1) = \frac{Z_2[x,a]}{(a^{n+1} + xa^n)}, \quad \text{where} \quad h^*(*) = H^*(RP^\infty) = Z_2[x].$$

2. Let $\rho': Z_2 \to O(n+1)$ be given by

$$T = \begin{bmatrix} -1 & & & \\ & -1 & & \\ & & \ddots & \\ & & & -1 \end{bmatrix},$$

then the induced action on RP^n is trivial, so we should get $S^\infty \times_{Z_2} (RP^n, \rho') = RP^\infty \times RP^n$, but the theorem gives us $Z_2[x,a]/(a+x)^{n+1}$, so what is wrong? The point is that if we let $(n+1): Z_2 \to O(n+1)$ be the trivial representation

$$T = \begin{bmatrix} 1 & & & \\ & 1 & & \\ & & \ddots & \\ & & & 1 \end{bmatrix}$$

and let $-1:Z_2 \to O(1)$ be the identity $T = [-1]$, then $\rho' = (n+1) \otimes (-1)$ and the standard Z_2-equivalence $P((n+1) \otimes (-1)) \xrightarrow{f} P((n+1))$ induced by the identity map $R^{n+1} \to R^{n+1}$ (which is not a Z_2-map $(R^{n+1},(n+1) \otimes (-1)) \to (R^{n+1},n+1))$ has the property that $h^*(f)(a) = a + x$, and of course $h^*(P(R^{n+1},n+1)) = Z_2[a,x]/(a^{n+1})$. We can be flippant about this and say that this is a case where the identity map does not induce the identity map in cohomology.

We now investigate the tangent bundle of $P(V)$ where V is a representation of G. The simplest case is $G = Z_2$, for if $\dim V = n+1$, then $V = r(-1) \oplus s\,1$, where $1:Z_2 \to O(1)$ is the trivial representation, $-1:Z_2 \to O(1)$ the nontrivial representation and of course $r+s = n+1$. Since $\tau(S(V)) \oplus 1$ is isomorphic to the G-bundle $S(V) \times V \xrightarrow{\pi_1} S(V)$, we have that $\tau(P(V)) \oplus 1$ is isomorphic to the bundle $S(V) \times_{Z_2} V \to P(V)$. Because $V = r(-1) \oplus s(1)$, this is a direct sum $r\hat{\lambda} \oplus s\lambda$, where λ and $\hat{\lambda}$ are the G-line bundles $\lambda:S(V) \times_{Z_2} (R,1) \to P(V)$ and $\lambda:S(V) \times_{Z_2} (R,-1) \to P(V)$. It remains to determine $w_1^{Z_2}(\lambda)$ and $w_1^{Z_2}(\lambda)$.

Lemma 7. We have $w_1^{Z_2}(\lambda) = a$, $w_1^{Z_2}(\hat{\lambda}) = a+x$, where

$$h^*(P(V)) = Z_2[x,a]/(q) \quad , \quad q = \sum_{k=0}^{n} w_k^{Z_2}(V)a^{n-k} \quad , \text{ and } a \in h^1(P(V)) \text{ is}$$

the class of the double cover $EZ_2 \times_{Z_2} S(V) \to EZ_2 \times_{Z_2} P(V)$.

Proof. We should distinguish between $G = Z_2$ and $O(1) = Z_2$. We have to show that the principal $O(1)$-bundle associated to $1 \times_{Z_2}:EZ_2 \times_{Z_2} (S(V) \times_{O(1)} R) \to EZ_2 \times_{Z_2} P(V)$ is the double cover corresponding to $EZ_2 \times_{Z_2} S(V)$, but this is immediate, for since Z_2 acts trivially on R, we have $EZ_2 \times_{Z_2} (S(V) \times_{O(1)} R) = (EZ_2 \times_{Z_2} S(V)) \times_{O(1)} R$, and the result follows: $w_1^{Z_2}(\lambda) = a$.

Now since $\hat{\lambda} = (-1) \otimes \lambda$, where $(-1):P(V) \times (R,-1) \to P(V)$ is the product bundle with the representation (-1) on R, we have $w_1^{Z_2}(\hat{\lambda}) = w_1^{Z_2}(-1) + w_1^{Z_2}(\lambda) = x + a$ (see the remarks before Corollary 5).

Note. The proof of Lemma 7 was helped a lot by some timely remarks of R. Stong, who also pointed out that it is easy to describe classifying maps for λ and $\hat{\lambda}$ in his model for $B(Z_2,O(1))$. Let $B(Z_2,O(1))$ be the space of lines in $R_-^\infty \oplus R_+^\infty$ where the element $T \neq 1$ in Z_2 acts by $T(\ldots,x_{-n},\ldots,x_{-1},x_0,x_1,\ldots,x_n,\ldots) = (\ldots,-x_{-n},\ldots,-x_{-1},-x_0,x_1,\ldots,x_n,\ldots)$. If $V = r(-1) \oplus s(1)$ then λ is classified by the inclusion $P(V) \subset B(Z_2,O(1))$ induced by $R_-^r \oplus R_+^s \subset R_-^\infty \oplus R_+^\infty$. The line bundle $\hat{\lambda}$ is classified by the map induced by the inclusion $R_-^r \oplus R_+^s \xrightarrow{f} R_-^s \oplus R_+^r \subset R_-^\infty \oplus R_+^\infty$, where $f(x,y) = (y,x)$ is not a Z_2-map, but $P(f)$ is, indeed $h^*(P(f))(a) = a+x$ (which by the way gives a new proof of Lemma 7).

Corollary 8. If $G = Z_2$, $V = r(-1) \oplus s(1)$, $\tau(V)$ the tangent bundle of $P(V)$ then

$$W^{Z_2}(\tau(V)) = (1 + a + x)^r(1 + a)^s ,$$

where $W^{Z_2}(\alpha) = 1 + w_1^{Z_2}(\alpha) + \ldots + w_n^{Z_2}(\alpha) + \ldots$ is the total Stiefel-Whitney class of α.

Example. The case $r = 1$ is of special importance since it turns out that $P((-1) \oplus n(1))$ together with a certain operation Γ on them furnish a set of multiplicative generators for the algebra of cobordism classes of Z_2-actions on closed smooth manifolds. Recall that $h^*(P((-1) \oplus n(1))) = Z_2[x,a]/(a^{n+1} + xa^n)$.

Proposition 9 (Bix). For each $n \geq 1$ the Z_2-manifold $P((-1) \oplus n(1))$ cannot be Z_2-equivariantly immersed into $m(1)$ for any natural number m.

Proof. It is sufficient to show that $(1 + a + x)^{-1}(1 + a)^{-n}$ has non-zero terms of arbitrarily high grading, but $(1+a+x)^{-1} = (\sum_{i=0}^{\infty} x^i)$ + (terms involving a^j, $j > 0$), $(1 + a)^{-1} = 1$ + (terms involving $a^j, j > 0$), so $(1 + a + x)^{-1}(1 + a)^{-n} = (\sum_{i=0}^{\infty} x^i)$ + (terms involving a^j, $j > 0$), so it has terms of arbitrarily high grading, as claimed.

Remark. More precise information is available: Bix [2] has shown that $P((-1) \oplus n(1))$ can be Z_2-equivariantly embedded in $n(-1) \oplus s(1)$ for some s, but it cannot be Z_2-equivariantly immersed up to cobordism in $(n-1)(-1) \oplus t(1)$ for any t. His results are even more general.

Regarding embeddings of $P((-1) \oplus n(1))$ into representation spaces of Z_2 the following result is useful.

Proposition 10 (Bix). If $P((-1) \oplus n(1)) \subset V$ is a Z_2-equivariant embedding, $\dim_R V = n+k$, $\nu : E \to P((-1) \oplus n(1))$ the normal bundle of the embedding, then $w_j^{Z_2}(\nu) = 0$ for $j > k$ and $w_k^{Z_2}(\nu) \in H^k(RP^\infty) \cdot 1$.

Remark. This result is the analogue of the result for ordinary Stiefel-Whitney classes that if $M^m \subset R^{m+k}$ then

$$\bar{w}_j = w_j(\nu) = 0 \quad \text{if } j \geq k.$$

So far we have only spoken about equivariant Stiefel-Whitney classes. The reader is invited to consult T. tom Dieck [8] for a definition of equivariant Stiefel-Whitney numbers. Recall that in the non-equivariant case a closed manifold M^m defines normal characteristic numbers: these are a homomorphism $H^m(BO;Z_2) \to Z_2$, or an element of $H_m(BO;Z_2)$. In the G-equivariant case a closed G-manifold M^m determines a homomorphism $X(M) : H^k(BO;Z_2) \to H^{k-m}(BG;Z_2)$ or an element of $H_*(BO;Z_2) \hat{\otimes} H^*(BG;Z_2)$ of degree m. Let us show how this operates by giving the proof of an early result of Bix (in the newer version of Bix [2] characteristic numbers are no longer used at this point and the new argument is remarkably simple).

Proposition 11 (Bix). The Z_2-manifold $P = ((-1) \oplus 2(1))$ does not equivariantly immerse in $k(1)$ for any k even up to Z_2-equivariant cobordism.

Proof. It suffices to show that $X(P)$ has terms of arbitrarily high algebraic degree in $b_1, b_2, \ldots$. To obtain $X(P)$ we conjugate the coefficient of a^2 in

$$(1 + b_1 a + b_2 a^2 + \ldots)^2 (1 + b_1(a + x) + b_2(a + x)^2 + \ldots)$$

$$= 1 + a^2 \sum_{i=1}^{\infty} b_i^2 x^{2i-2})(1 + (a + x) \sum_{j=1}^{\infty} b_j (a + x)^{j-1})$$

$$= 1 + a^2 \sum_{i=1}^{\infty} b_i^2 x^{2i-2} + (a + x) \sum_{j=1}^{\infty} b_j (a + x)^{j-1}$$

since $a^2(a + x) = 0$. So $X(P) = \sum_{i=1}^{\infty} \bar{b}_i^2 x^{2i-2} + \sum_{j=1}^{\infty} \bar{b}_{2j} x^{2j-2}$, where the bar denotes conjugation. Let $\Delta(z) = \sum_{i=0}^{\infty} \Delta_i(z) s^i$ as in Lecture 1, then

$$\Delta\chi(P) = \sum_{i=1}^{\infty} [\Delta\bar{b}_{2i} + (\Delta\bar{b}_i)^2]x^{2i-2}$$

$$= \sum_{i=1}^{\infty} [\sum_{j=0}^{2i} \bar{b}_j s^{2i-j} + (\sum_{j=0}^{i} \bar{b}_j s^{i-j})^2]x^{2i-2}$$

$$= \sum_{i=1}^{\infty} (\bar{b}_1 s^{2i-1} + \text{terms in } s^j \text{ with } 0 \le j \le 2i-2)x^{2i-2}$$

and thus $\Delta\chi(P)$ has terms involving arbitrarily large powers of i, so $\chi(P)$ has terms of arbitrarily large algebraic degree in $b_1,\ldots,b_n,\ldots$, as claimed.

References

[1] J.F. Adams, On the nonexistence of elements of Hopf invariant one, Annals. of Math. 72 (1960), 20-104.

[2] M. Bix, Ph.D. Thesis, University of Chicago, 1974.

[3] J.M. Boardman, Stable homotopy theory, University of Warwick, Coventry, 1965.

[4] R.L.W. Brown, Imbeddings, immersions, and cobordism of differentiable manifolds, Bulletin A.M.S. 76 (1970), 763-766.

[5] ____________, Immersions and embeddings up to cobordism, Canadian J. Math. 13 (1971), 1102-1115.

[6] T. tom Dieck, Faserbündel mit Gruppenoperation, Arch. Math. 20 (1969), 136-143.

[7] ____________, Bordism of G-manifolds and integrality theorems, Topology 9 (1970), 345-358.

[8] ____________, Characteristic numbers of G-manifolds, Inventiones Math. 13 (1971), 213-224.

[9] A. Dold, Erzeugende der Thomschen Algebra N_*, Math. Zeitschrift 65 (1956), 24-35.

[10] ________, Partitions of unity in the theory of fibrations, Annals of Math. 78 (1963), 223-255.

[11] S. Gitler, Immersion and embedding of manifolds, Proceedings of Symposia in Pure Mathematics 22 (1971), 87-96.

[12] M.W. Hirsch, Immersions of manifolds, Transactions A.M.S. 93 (1959), 242-276.

[13] I. Kozma, Witt vectors and generators for cobordism (to appear).

[14] A. Liulevicius, A proof of Thom's theorem, Commentarii Math. Helv. 37 (1962), 121-131.

[15] ______________, Homology comodules, Transactions A.M.S. 134 (1968), 375-382.

[16] ______________, Immersions up to cobordism. Preprint, University of Chicago, November 1972.

[17] J.W. Milnor, The Steenrod algebra and its dual, Annals of Math. 67 (1958), 150-171.

[18] R.E. Stong, Unoriented bordism and actions of finite groups, Memoirs of the A.M.S. 103 (1970).

[19] R. Thom, Quelques propriétés globales des variétés différentiables, Commentarii Math. Helv. 28 (1954), 17-86.

[20] R. VandeVelde, An algebra of symmetric functions and applications, Ph.D. Thesis, University of Chicago, 1967.

[21] R. Wells, Cobordism groups of immersions, Topology 5 (1966), 281-294.

TEMPORAL EVOLUTION OF CATASTROPHES

René Thom

Institut des Hautes Etudes Scientifiques, Bures-sur-Yvette

First Lecture

I will devote this first talk to a general presentation of catastrophe theory and to the basic purpose of this theory, as I see it.

Let us consider any natural process taking place in some domain $D \times T$ of space-time. Catastrophe theory is basically interested in morphology; i.e. in discontinuities in the qualitative properties of the medium. As soon as one speaks about forms, about morphology, or about a phenomenon in general, one has to accept the idea that the form can be defined as a qualitative discontinuity inside some specific medium. The topological space on which the form is defined will be called the substrate space.

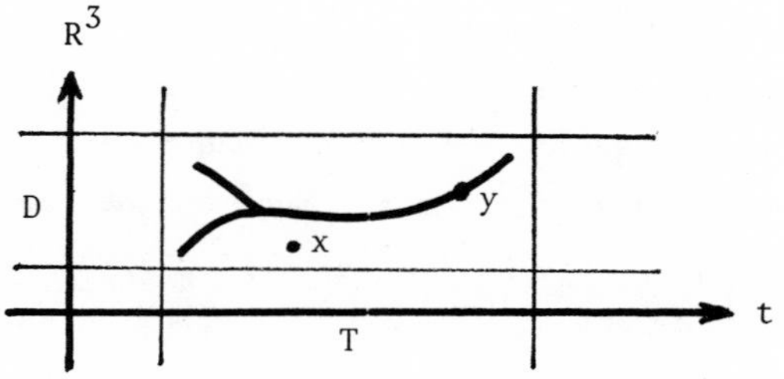

To define the form, to define the morphology, I will use the following idealization. A point x inside the substrate space will be said to be a regular point of the morphology if all nearby points have the same qualitative properties as x ; i.e. if there exists an open ball B_x around x such that for any point x' of B_x the medium has the same qualitative properties in x' as in x. This definition is, of course, very vague but I think one has to start with a definition even if it is not entirely formalizable. Using this definition one sees immediately that the set of regular points is an open set in the substrate space, and we shall be interested in the complementary set of this open set, which is called the catastrophe set. Given any point y belonging to the catastrophe set, then, in any ball around y, there are points which do not have the same qualitative property as y; one gets some abrupt change of the properties of the medium around the point y. One sees from this general definition of the set of catastrophe points for any natural process that the catastrophe set - as the complement of an open set - is closed, and that the data of the morphology always imply the description of its catastrophe set.

The catastrophe set itself, of course, does not give the full phenomenology of the process, because it does not take into account the continuous variations of the properties of the medium. (If, for instance, you consider something like temperature or colour, things may vary continuously inside the medium and this will not be described by a catastrophe in general). Nevertheless, I think we must take this definition as a starting point for any morphological study in general. One might object that this definition is not very precise because it depends on the tools which are used to examine the process. That is quite true. If you look at the neighbourhood of a regular point with a microscope you could well see a lot of things which you would not be able to see with the naked eye; so with a microscope you would say that this point is a catastrophe point. I agree with these difficulties, they are more or less inherent to any kind of phenomenological theory, but let us say that the distinction between regular and catastrophe points is associated to a specific way of looking at phenomena. Much can be said in favour of this idea. In perception theory, for instance, psychologists have used since a long time the notion of "figure and ground". When one sees an object, it is separated from the ground by a specific boundary curve. This boundary curve has to be looked at as a set of catastrophe points. (In fact this notion may be considered as the fundamental motivation for general topology and for the notion of continuity in general).

We shall now consider a situation in which the morphology is given by an experiment. By that I mean that it is possible to prepare in the domain D some specific system in a specific physical state such that if we prepare this system in this state at the time $t = 0$ then we are able to look at the evolution of the system while time increases. It might happen that during this evolution some qualitative discontinuities will appear inside the medium.

We have to be a bit more precise. What does it mean to say that we prepare a certain physical state inside the domain? First of all, the domain is separated from the surrounding universe by a kind of wall, in general it is a box (we always do experiments in boxes), and we fill this box with specific chemicals, substances, living beings, or anything you like, according to a very specific preparation procedure. The preparation procedure is a text written in ordinary language which describes all the operations which are needed in order to prepare the system in a given state α .

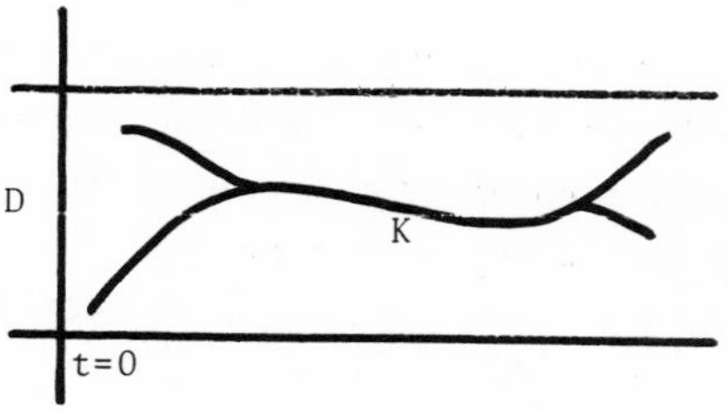

state α
catastrophe set K

Now we shall say that the morphology which we get out of state α is structurally stable (or, to use a more sophisticated language, is given by a morphogenetic field) if the following condition is satisfied. Suppose that somebody else makes the same experiment; i.e. he takes another box D' which is equivalent to D by some linear Euclidean transformation E, and then prepares the same system in the same state α .

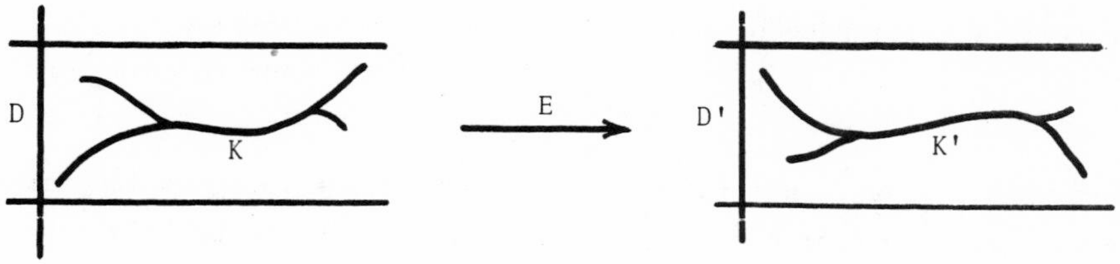

The morphology will be said to be structurally stable if the new catastrophe set K' is obtained from the old one by applying the Euclidean transformation E to K, followed by some ε-homeomorphism h

$$K' = h_\varepsilon \circ E(K), \quad \text{where} \quad d(h_\varepsilon(x), x) < \varepsilon.$$

This notion of structural stability is, of course, very natural, because of the fact that, implicitly, in all sciences we believe in the regularity of phenomena. We know that the universe is not as variable as one could expect and there are many local situations which lead to the same qualitative results. In fact, it is well known that any experimental fact has to be reproducible, and this requirement of reproducibility is exactly the same as that of structural stability. So, if one sticks to this requirement of reproducibility for experimental scientific facts, one actually is led to the notion of structurally stable morphologies.

I mentioned already that one can use the notion of a morphogenetic field to describe this more precisely. This can be done in the following way. The morphogenetic field is defined by a universal substrate space $\tilde{U}$. In this universal substrate space you have the universal catastrophe set $\tilde{K} \subset \tilde{U}$. To say that the morphology is given by this morphogenetic field means that there always exists a mapping

$$\Phi: D \times T \longrightarrow \tilde{U}$$

(in general a homeomorphism or, in many situations, a diffeomorphism) such that the catastrophe set K of the experiment is the counter-image under Φ of the universal catastrophe set

$$K = \Phi^{-1}(\tilde{K}).$$

This is the abstract definition of a morphogenetic field. In some cases it is interesting to look at the evolution in time, so we might suppose that the time axis is also in the universal space and one requires then that Φ be compatible with the projection onto the time axis; i.e. that Φ transports a hyperspace of simultaneity into a hyperspace of simultaneity. But this is a refinement which I don't want to consider here.

When we are given a natural morphology, the first thing to look at is whether one can describe all the morphologies one meets by only a finite number of morphogenetic fields. This suggests the notion of a morphology of finite type. We say that the system of morphologies given by experiment or by observation is of finite type if there is a finite number of morphogenetic fields, say $\tilde{U}_i \supset \tilde{K}_i$, such that for any experimental morphology one may cover the catastrophe set with a locally finite covering, so that for each of these local charts V_i we have $V_i \cap K = \tilde{K}_i \subset \tilde{U}_i$.

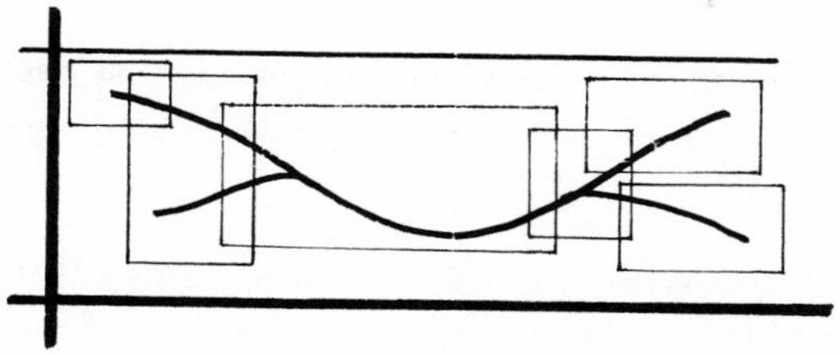

Consider a mathematical example. Look at the ordinary cut locus of a Riemannian surface; i.e. a two-dimensional manifold with a Riemann metric. Taking any point p, one defines the cut locus as the set of points q where one may draw two geodesics which are minimal from p to q, or one geodesic which is not singular. One can show that for almost all points p the cut locus has only specific types of singularities, namely the following three types: the regular differentiable arc, or the triple point, or the free vertex.

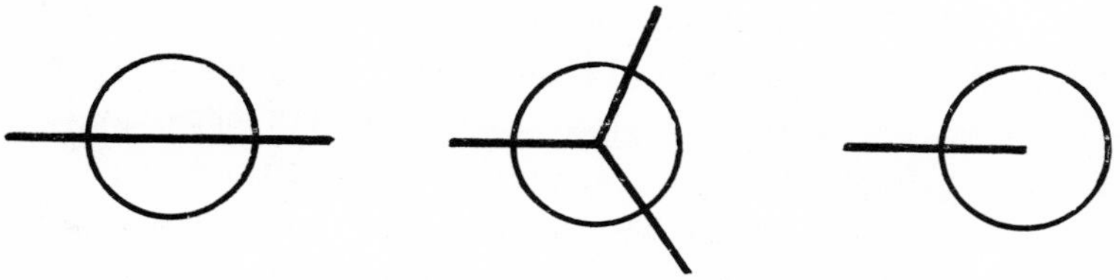

One can cover any cut locus of the general type by these three types of local singularities. So, in this case, it is possible to describe almost any cut locus on a Riemannian surface by a finite alphabet of singularities.

The same is true, for instance, for the critical locus of the critical set of a mapping for almost all mappings of a manifold into another. The set of critical points and the set of critical values are in general only of a finite number of local topological types. This indicates why this notion of a morphology of finite type has a very strong mathematical motivation.

Let us consider another example which is, of course, very useful, namely a linguistic morphology. An ordinary spoken text can be decomposed into sentences, the sentences into words, then each word into syllables, and each syllable into isolated elements which are called phonems (in a written text, the phonems are just ordinary letters). If you look at the linguistic morphology as a one-dimensional morphology with substrate space the time axis, then each phonem can be considered as a morphogenetic field and then the linguistic morphology is nothing but a sequence of such morphogenetic fields. In any language we have only a finite number of phonems, called the phonetic system of the language. So the linguistic morphology is to be considered as a morphology of finite type.

In the same way in biology, for instance, (- although this has never been made explicit by biologists, because they are not interested in this type of abstract thinking -) any organism can be decomposed into organs, each organ being formed by cells of specific differentiation types (e.g. liver, lung, bones, muscles, etc.). Now it happens that if you subdivide each organ into cells of the same differentiation type then you get only a finite number of types of cells, and, moreover, the number of characteristic positions of each type of tissue with respect to the surrounding tissue is always a finite number, too. So here again, looking at the biological morphology of an organism, we are dealing with a morphology of finite type.

There is a kind of philosophical problem which arises when we deal with sciences which are not experimental but only purely observational. There are scientific disciplines, such as astronomy, paleontology, geomorphology, etc., in which it is practically impossible to perform experiments, because, for instance in astronomy, of the distance and size of the stellar objects. Paleontology deals with the past and there is absolutely no way of experimenting with past events; similarly geomorphology is a purely observational science and does not pretend to make experiments. For all these reasons there are many morphological theories which are purely observational. So, for these types of disciplines it seems to be impossible to define the notion of a morphogenetic field. Nevertheless, even there one sees many local accidents which occur very frequently with a great regularity, and in that case one may suspect that they are locally structurally stable. One may assume that these accidents, which occur very regularly, are given by a morphogenetic field. This is a very natural assumption to make. So, even for purely observational theories, you can use this formalization, and can see whether or not the morphology is of finite type.

There is another very interesting notion, namely that of the hierarchical levels of organization. It might happen that, if you look at a given experimental morphology, some aggregate of local morphogenetic fields exhibit a very strong stability and a very strong frequency. It might well occur that some of these aggregates can be stabilized by imposing constraints in the preparation procedure of the state. As a typical example, let us consider living beings. If you want to prepare an animal of a given species the only way of preparing this animal up to now is to look in nature or some other laboratories for animals of the same species and to get two animals of opposite sexes into a convenient medium and then they will reproduce. This is a very tautological way of preparing animals, but one has never been able to find anything better up to now. In that case the constraints at the initial time for an animal of a given species S is precisely to have these animals at the initial time.

More generally, one can write down the specific constraints which will ensure that some aggregate of elementary morphogenetic fields will be stabilized and behave as global morphogenetic fields. In that case we shall speak about conditional morphogenetic fields. One might have several successive levels of conditional morphogenetic fields; a linguistic morphology is a very typical example of this, because it has a complete hierarchy of levels from the phonems to the syllables and to the words, sentences, and to the complete text. Each of these aggregations of elementary morphogenetic fields can be stabilized by subjecting the initial data to some specific constraints.

Now, what is the general purpose of any morphological theory? The answer to this question consists of several parts:

A. Find out whether a morphology is of finite type and describe the "alphabet" (i.e. the set of elementary morphogenetic fields).

B. Find the conditional morphogenetic fields by describing the constraints and the initial data which stabilize them, and, moreover, describe the complex conditional morphogenetic field as an aggregation of elementary morphogenetic fields; i.e. the structure of the conditional morphogenetic field (which is the geometric description of its decomposition into elementary morphogenetic fields). In the case of a one-dimensional morphology, such as we have for a linguistic morphology, the description of this structure is, of course, quite easy, because each word or each syllable is a concatenation of elementary phonems, so the description of the structure is nothing but a description of this concatenation of elementary symbols. So, a one-dimensional morphology of finite type can always be completely algebraized; any such morphology is a word of the free monoid generated by the finitely many elements of the alphabet. So, for one-dimensional morphologies, we have a complete algebraization. For multidimensional morphologies the situation is much more complex, because when one has a complex object, then, in order to describe its structure, one has to give a covering and prescribe the position of each element of the covering with respect to the others; so one needs something like the nerve of the covering. This problem of algebraization in the multi-dimensional case is the basic difficulty for the study of multidimensional morphologies.

C. Find the structure describing the hierarchical levels of organization (this is the notion which is not very clearly defined). By this I mean essentially that we have to be able to define a conditional morphogenetic field of level j, say, to order the successive levels, and to find the structure of an aggregation of a field of level $j + 1$ as a (spatial) aggregate of fields of level j. For instance, if you look at phonems and syllables, you can say that each syllable is composed of a very small number of phonems. (In many languages one has a kind of canonical form for syllables in terms of one central vowel and some consonants, and there are very specific rules in each language to describe the specific types of syllables which are admissible). In the same way a sentence can always be reduced to words, and this is exactly the purpose of the so-called syntax or grammar, to describe how a sentence can be composed from words; i.e. to describe specific rules which have to be satisfied in order for a sentence to be "correct". Here again one has very canonical rules. Consider, for example, the sentence "the cat catches the mouse". Its morphology is given by the following diagram (where P = proposition; SS = syntactic subject, SV = syntactic verb, SO = syntactic object; A = article, N = noun, V = verb).

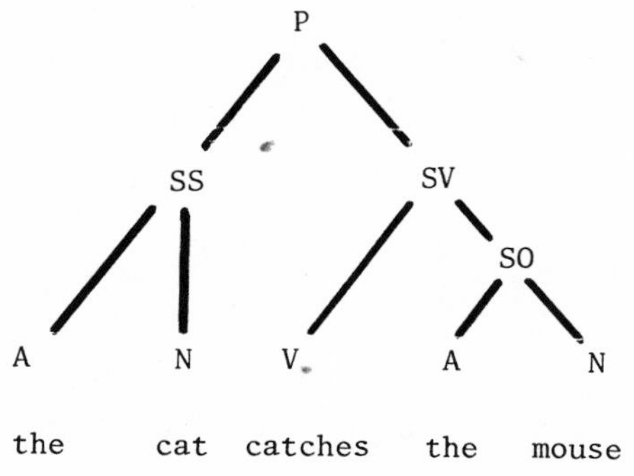

The grammatical structure of this sentence is completely described by the above tree. This means that if we consider the levels of sentences and of words, one is able to map the morphology of the sentence into an abstract morphology whose elements are these symbols noun, article, verb, etc. So the notion of hierarchical level can be described as follows. We have an abstract morphology M of finite type, and a homomorphism from the conditional field of level $j + 1$ into the objects of this abstract morphology M

$$(\quad)_{j+1} \longrightarrow M$$

such that each image element is decomposed into an aggregate of elementary objects of the alphabet of M, and each of these elementary objects being itself the image under this homomorphism of an element of level j.

For instance, coming back to the level of syllables and phonems, we have specific types of syllables, where the basic symbols of the abstract morphology are V (vowel) and C (consonant); again there are only finitely many admissible symbols for describing the structure of syllables. In the same way, for an elementary sentence, we have a finite number of aggregates of words, given by the last row of such a tree, and these correspond to the abstract morphology M. Strangely enough, the elements of this abstract morphology are frequently called "functions" (of course, not in the mathematical but in the grammatical sense).

In the same way, in biology, if one looks at an organism, a living being, then its body can be decomposed into organs and in a purely abstract way, one could define a kind of idealized body, and in this idealized body successive idealized organs, in such a way as to be able to map any living organism into this abstract morphology (this "abstract animal"). This was the subject of a classical discussion in the French Academy of Sciences around 1830 between Cuvier and Geoffroy Saint-Hilaire. The latter claimed that there should exist a universal scheme for all kinds of animals and that one could describe the structure of all animals by mapping them into this general abstract structure. Of course, for vertebrates this is an idea which is very easy to defend (I think this is now generally admitted), but when you claim this in its universal form, for vertebrates, anthropodes, insects etc., the situation becomes much more complicated. Now most biologists believe with Cuvier that the idea of a universal pattern for all animals is a kind of dream which one should not continue to have. But, nevertheless, I believe that it is really a basic problem of theoretical biology to understand whether this general pattern does exist or not.

D. Finally, form all possible experimental morphologies; i.e. form the "corpus of all observable morphologies". This is the role of the experimenter in general. Of course, this corpus is, in general, infinite, you can never be sure that you can obtain all possible morphologies in a given science, but for practical reasons you have to stop with a finite corpus and you stop by supposing that the new data won't change very much the results you got from the preceding corpus.

This is what is called the stage of description. In any morphological discipline the first thing to do is to get a complete description, and this is the programme of all descriptive sciences. Now, after having such a description, there arises the problem of understanding what to do with the description and, of course, most people will ask for an explanation next. Here is a very difficult and delicate problem of the philosophy of science: to understand what is the relation between description and explanation.

Many people would say that what is important is explanation. I think that most scientists do not have a very clearcut idea about this. In many situations, for example when I am speaking about catastrophe theory, biologists will say: "Well, what you give is just a description, not an explanation, and what we want is an explanation! In anatomy textbooks one describes the morphology, in embryological textbooks one describes the embryological development, so what is the interest of putting new things into the description if it is not giving an explanation?". And they will point out that an explanation in biology should always be given in terms of lower level elements; i.e. in terms of cells, or, eventually, in terms of molecules.

Let me say that there are two basic attitudes or approaches. First there is the "reductionist's approach", which states that, in order to explain, one has to look at the way the given morphology is produced (its "cause") and then to pass to the lower elements, and from the interactions between these lower level elements one should reconstruct the higher order morphology. So the reductionist's attitude is basically an atomistic approach which supposes that there are elementary elements as "atoms" and that one can give a differential system describing the evolution of these elementary elements and their interactions. The basic model for this is the one in which each atom is described by its position and velocity, and we write down the law of interaction, the potential of interactions of these atoms being a huge differential system

$$\frac{\partial x_i}{\partial t} = X_i(x_j) \quad ,$$

and by integrating this differential system we should be able to reconstruct the full evolution of the system. This reductionist's atomistic approach is, of course, theoretically very satisfactory, because it is a complete reduction to the computation of a quantitative model, and if it could be done it would certainly give a complete and positive answer to our problem. But, unfortunately, it is in general unmanageable, because of several reasons:

(1) In general, the elementary elements are not known in many situations.

(2) Even if we do know what the elementary elements are, we have to describe their states and their interactions, which may be very complicated. Even for molecules, for instance, the interaction between nearby molecules is a very difficult problem for which one does not have a complete quantitative solution. (This is a point molecular biologists are not eager to accept, but I think it is true. Even in the case of only two molecules, a correct quantitative description of all their interactions may be too complicated a problem in general).

(3) The high dimensionality of the system makes it impossible to find a

quantitative solution (e.g. in a perfect gas, you have the Avogadro number of elementary elements in the unit mass of the substance, giving rise to a problem involving 10^{23} dimensions, which in general, of course, is impossible to integrate).

(4) Moreover, there is another fundamental difficulty in the reductionist's approach, namely, if we stick to the idea of having a quantitative model then we should get analytic functions on the right hand side of the above equations, in order to get analytic solutions (because analytic solutions are practically the only ones which one can compute). So one has, in particular, to consider all solutions continuous, which excludes discontinuities. But in morphological theories one is exactly interested in discontinuities, and that is an additional reason why the usual quantitative modelling is not appropriate for morphological theories.

There is another approach possible which I shall call the structuralist's approach (which is the one I advocate), and which is - in some sense - less ambitious than the reductionist's approach. To explain what I mean by the structuralist's approach, let me give an example. Look as Newton's gravitation law

$$E = \mu \frac{mm'}{r^2}$$

(which is usually considered a standard example of "explanation"). This law explains a lot of things, for example the motion of celestial bodies, the gravitation on earth, etc. What is Newton's law of gravitation, is it description or explanation? Many will say that it is not really an explanation, because nobody understands up to now the true nature of gravitation, nobody understands really why two masses attract each other and what is the underlying mechanism of this attraction. So one should probably say that Newton's law of gravitation is not really an explanation, but is only a more complete description, and the progress obtained by Newton's law is basically due to the fact that it involves an enormous simplification of description. Before, you had lots of data concerning the motion of planets, Kepler's laws, gravitation on earth, etc.; you had a lot of seemingly unrelated phenomena. Using Newton's law, all these phenomena are fitted together into a single scheme and can be deduced from this single scheme very easily by purely mathematical means.

So I think this is the way one has to look at things: the purpose of any scientific morphological theory in general is to get description on an economical basis, to try to find the most economical way of organizing the description and to reduce the arbitrariness of the description. For instance, in formal language theory, one wants to find an axiomatization of all given experimental morphologies and one wants to write down a finite number of axioms, such that by combining these axioms one gets a complete list of the words which appear experimentally. I think this is a very basic example; one should always try to generate axiomatically all the given morphologies. The notion of generating things axiomatically is, of course, a very standard notion in algebra, in the one-dimensional morphologies, because here one deals with a situation in which one is in a free monoid, so one has the full algebraic theory at one's disposal. When one deals with multi-dimensional morphologies, however, the notion of "generation" is not defined, and the notion of "operation" on a multi-dimensional morphology is not defined, so one has to define it. It is the purpose of catastrophe theory to do precisely that; catastrophe theory really wants to describe the most obvious way of aggregation of elementary morphological accidents in multi-dimensional situations, so that one is able to develop an algebraic theory dealing with these multi-dimensional morphologies. Roughly speaking, it is the homologue to the standard operations of arithmetic, except that one deals with morphologies and not with numbers.

Second Lecture

In my first lecture I described the general programme of a morphological theory and I explained that, in order to find an algebraization of local morphological accidents, the best tool - or perhaps the only tool - known up to now is the theory of catastrophes. So I shall try to explain a model of catastrophe theory, and I will do that using implicitly the example of biology, because I want to say more about biology in my third lecture anyway.

So let us suppose we consider an egg which is dividing and developing into an embryo. Then the egg is considered to consist of cells and one might suppose that one may parametrize the state of a cell by a certain very large number of parameters. So, given any cell, we assume that it is possible to parametrize the possible biochemical metabolic, biological, and any kind of states, for any kind of property, by the values of a certain number of parameters $y_1, y_2, \ldots y_N$, N large. Then, to any cell c_i, there are associated the values of the corresponding parameters for this cell, $y_j(c_i)$. In this way one gets for any cell of the organism a corresponding point y in this huge Euclidean space R^N. Let us agree that this huge Euclidean space will be called the space of internal variables, because it describes all the internal properties of the cell, and I will also consider the usual space-time R^4 as base space, called the space of external variables.

The main problem one deals with in embryology is to describe the spatio-temporal evolution of the egg with respect to its cellular differentiation, i.e. each cell will develop into a specific type of cell (e.g. a bone cell, a muscle cell, a blood cell, etc.) and according to this specialization and differentiation the full organism will develop into a kind of stratified structure, and the problem really is to find first a description of this process and, later on, to find some kind of "explanation" for this whole process.

The basic scheme of catastrophe theory is as follows: we assume that the local state of a cell is always obtained through an optimization principle; i.e. one is given a priori, a real-valued function $S(y;x)$, where y is the internal and x the external variable, and we assume that this function depends smoothly on the coordinates y and x such that, for any point $x \in R^4$ of the external space, the cells tend to maximize (or minimize) in the corresponding fibre the given function S. S could be called a "local entropy".

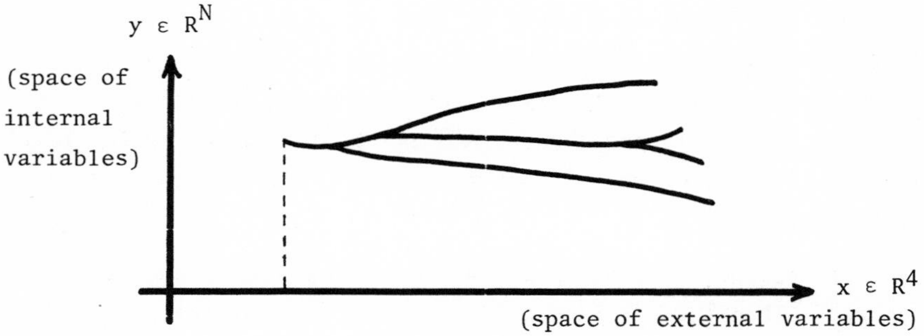

So the basic idea is that, for any poiny of the base space, the corresponding cells lying above this point will tend to maximize this function. So in order to get the possible equilibrium position for the cells, I will consider the first derivative

$$\frac{\partial S(y;x)}{\partial y} = 0$$

and, from the critical points of this S-function, find out the maxima; then the cells will tend to aggregate towards these maxima, and these maxima will define the local equilibrium position of the corresponding cells. You see that when you follow the evolution of a cell through time then each cell is represented by a point, so you will get a system of "curves".

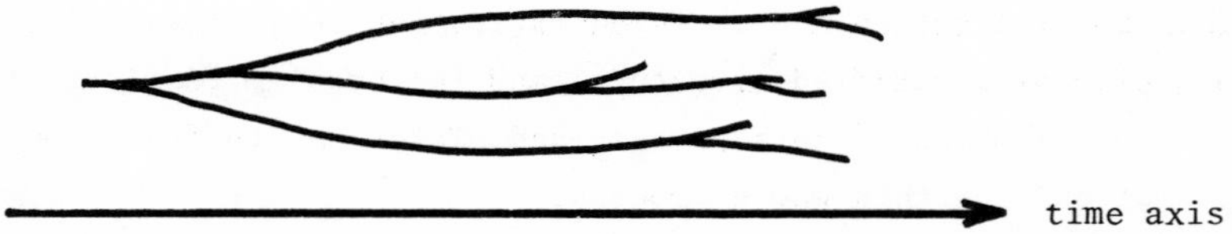

If the cell divides, you get a process with the very natural assumption that the two daughter cells are "very near" to the original parent cell. So, heredity is a continuous process with respect to the topology defined by these parameters. What one gets is basically that, for each time t, one has in the product space of R^3 with the space R^N of internal variables a cloud of points describing all possible states of the cells, and the problem is to find a formalism which describes the evolution of this cloud of points.

It is now very convenient to suppose that this cloud of points comes from a continuous map of a cell (the system of cells forms a sort of "discretization" of a smooth mapping) so that we have a mapping

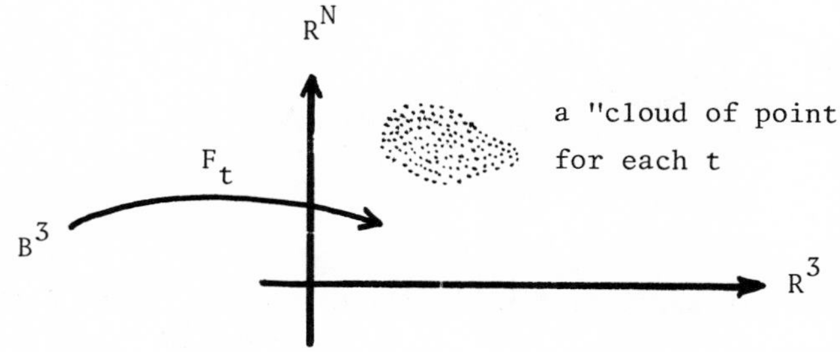

- (B^3 is a 3-dimensional disc or cell, and is biologically a cell, too!) - which maps B^3 into this product space $R^N \times R^3$; and we assume then, by smoothing and intercollating the whole situation, that at each point the image of the map tends to satisfy this maximizing condition, this optimality principle.

So we have a type of "local optimality principle" as the first axiom of elementary catastrophe theory. The second axiom is the axiom of "genericity" of the S-function. These are the two basic axioms of elementary catastrophe theory. I explained already the meaning of the first axiom, namely, that the image of the map F_t tends to occupy the maximum of the S-function. The genericity of the S-function is slightly more difficult to define mathematically. Consider the function $S(y;x)$ as a mapping from the base space R^4 to the space $C^\infty(R^N,R)$ of C^∞-functions from R^N into R, by associating $x_o \in R^4$ with $S(y;x_o) \in C^\infty(R^N,R)$; denote this map by

$$\Phi : R^4 \longrightarrow C^\infty(R^N,R).$$

In this function space there is a certain kind of object which I call the "canonical stratification" $\sum$ of the function space which, in some sense, describes the topological and differentiable properties of the function Φ. I recall briefly the definition. In the function space you have first $\sum^o$ which is the open dense set of "excellent" functions (an excellent function is one whose critical points are all non-degenerate and whose critical values are all distinct; as you probably know, such functions are structurally stable with respect to perturbations). An example in one variable is the following

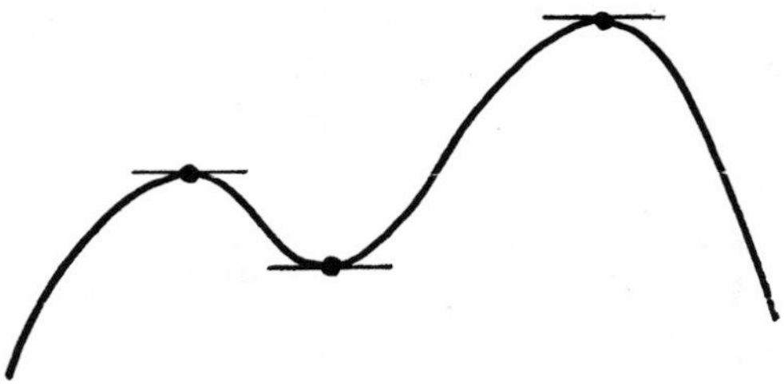

with 3 critical points and 3 distinct critical values. This example-function is structurally stable; i.e. if you perturb it slightly, then the new function will be differentiably equivalent to the original one, which means that you can transform the original function into the new one by a diffeomorphism of the total space. So we have an open dense set, and this is what is called the stratum of codimension 0 in the function space $C^\infty(R^N,R)$. Now you consider the complementary set $\sum^1 = \sum - \sum^o$; the regular part of this complementary set consists of two submanifolds of codimension 1, namely $\sum^1_\alpha$ and $\sum^1_\beta$. $\sum^1_\alpha$ consists of all functions which satisfy the condition that all critical points are non-degenerate, except one of type $y^3 + \sum y_0 \pm \sum y_1^2$, and all critical values distinct; whereas $\sum^1_\beta$ consists of those functions for which all critical points are non-degenerate and all critical values distinct, except two of them equal.

As one sees from these definitions, both $\sum^1_\alpha$ and $\sum^1_\beta$ are submanifolds of codimension one of the function space, because only one condition has to be satisfied; in this way we get nicely embedded submanifolds. Now one may continue this process and form an open and dense set in $\sum^1$ to get a $\sum^2$ consisting of things of codimension two, etc. I omit the full description of this. Thereby one gets a decomposition of the function space as a union of manifolds and we are interested only in those manifolds which are of finite codimension. The infinite codimensional strata are of interest to analysts but not to topologists, I would say, and the genericity assumption for the S-function means now that the above mapping $\Phi : R^4 \rightarrow C^\infty(R^N,R)$ is transversal, in general position, to the strata of this stratification.

This may look a bit abstract at first as a definition, but I think it is perhaps the easiest way of giving the meaning of this notion of genericity. Intuitively it means that all singularities of this function are the least degenerate possible with respect to variation of the base point. The genericity assumption is in fact an assumption of general position.

Now, what is the consequence of these two assumptions, which are in fact very weak? We are interested not in what happens in that big space (because nobody knows enough about the nature of the space of internal variables), but we are interested in the projection of the graph onto the base space, and these two axioms imply some conditions about the nature of the singularities of the image space of the critical value set, when we project this graph onto the base space. I will try now to explain what kind of conditions one gets.

Given any point x, the corresponding equilibrium positions are the maxima of the S-function. Instead of speaking about the entropy function S, I would like to speak about a potential function V

$$S(y;x) = -V(y;z) = -\text{grad}_y\ V.$$

The maxima of the S-function are now the minima of the potential function; one assumes that the local cells are moved by a dynamics which is defined as $-\text{grad}_y\ V$; so we consider the minimum of the potential function.

What is the relation between this model and the idealization I gave in the first lecture? What will be the state of a point of the space of external variables? The state will be the specific minimum of the corresponding potential function. In general, this minimum will be a non-degenerate quadratic minimum, hence it will be structurally stable. Now, if I have chosen such a minimum, say μ, of x then this minimum will stay for a sufficiently small neighbourhood of the point x, by the stability of a non-degenerate quadratic minimum. So I may extend the corresponding equilibrium position on such a neighbourhood. If the local state of this is really described by such a branch of local minima I will say that x is a regular point of the morphology; if, however, we get a discontinuity, a jump from one minimum to another for a specific point x' of the base space, then x' is said to be a catastrophe point of the morphology. So we have defined a catastrophe point in the following way: we have a section σ for a definite value in the minimum set of the global potential function; for regular points this section σ is continuous, and for catastrophe points σ is locally discontinuous.

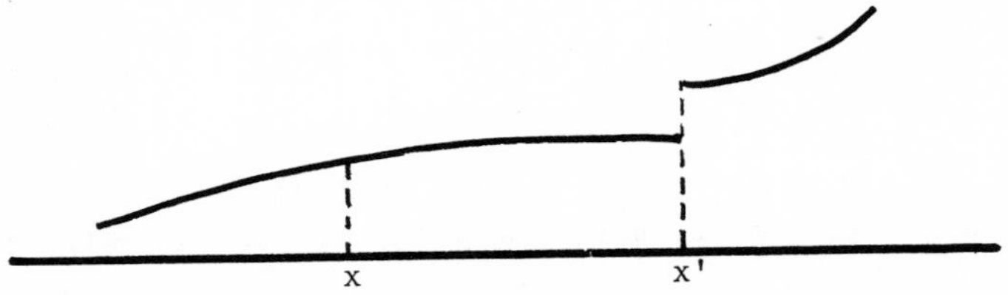

regular point catastrophe point
of the morphology

Now one might ask the following question, which is really a fundamental problem of this whole theory: suppose I have chosen the point x_o in the base space and I consider the corresponding potential function $V(y;x_o)$, and suppose this function has two minima, namely at μ_o and μ_1; to fix ideas let us furthermore assume that we have only one parameter y for the space of internal variables. The question is: which is the dominating minimum at x_o?

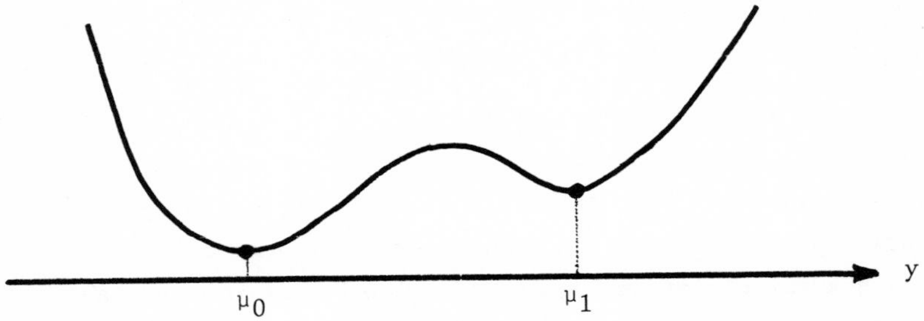

Here we deal, of course, with a very concrete physical or biological problem, hence there is no hope of getting a mathematical solution to this type of question. The only hope is to get a mathematical answer without knowing whether this mathematical answer will have validity for applying this model to concrete situations. The best mathematical answer, the easiest mathematical answer, is the so-called Maxwell convention which says that the dominating regime is the minimum of lowest value; in the above figure, for instance, the minimum at μ_0 is lower than that at μ_1, so it will " dominate" the other. If we accept this convention,then when should we get catastrophes? We should get catastrophes for positions for which the absolute minimum either ceases to be non-degenerate or when it is attained in at least two points. This defines, in our stratification Σ, a big set Mx (Maxwell set); it is a substratified set of the stratification. So, to define Mx explicitly, a function V belongs to Mx if and only if either the absolute minimum of V is degenerate, or the absolute minimum of V is attained in at least two points.

Hence the complementary set of the Maxwell set Mx is the set where the absolute minimum is simple and non-degenerate. This is obviously an open and dense condition, so Mx is a closed and nowhere dense set in the space C^∞ of functions. Then the catastrophe set of our process will be given by the counterimage under the mapping Φ of the Maxwell set Mx,

$$\Phi^{-1}(Mx) = \text{catastrophe set.}$$

I shall now give the simplest example of a catastrophe (and I have to apologize to those of you who have heard this classical story already). The simplest example of a catastrophe is the so-called <u>Riemann-Hugoniot catastrophe</u>. (Actually this is not exactly true: the simplest possible type of catastrophe point is, of course, given by the strata of codimension one of the Maxwell set; i.e. the strata one gets by taking functions whose absolute minimum is reached in precisely two points which are non-degenerate minima. This defines a nicely embedded submanifold of codimension one, i.e. a hypersurface, in the function space. So, if your mapping Φ is a smooth mapping which is transversal on this hypersurface then the counter image is just a nice smooth hypersurface; so this is the simplest type of catastrophe, called the "conflict stratum". It is more

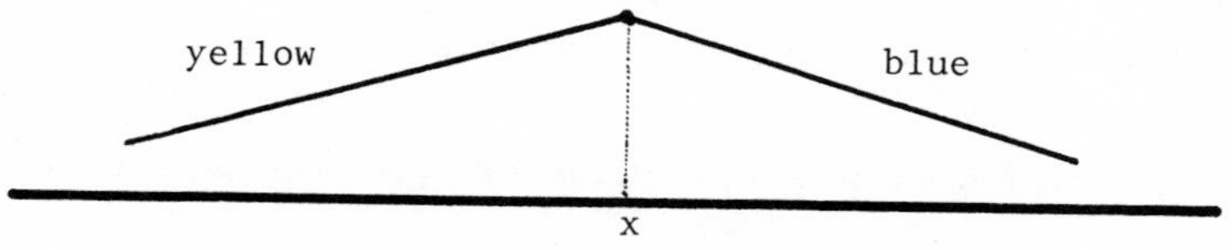

or less the boundary between two conflicting regimes, say yellow and blue). The next simplest type of catastrophe is the Riemann-Hugoniot catastrophe, so let me describe this now.

This catastrophe is related to a strata of codimension two in the function space, so I have to quickly describe this strata of codimension two. We first have $\sum_\alpha^2$, the set of functions which admit one minimum of type $y_o^4 \pm \sum y^2$, and we have some other strata of codimension two which are of lesser interest, namely the set $\sum_\beta^2$ of those functions for which the absolute minimum is reached in three points (non-degenerate), and also, for instance, the stratum $\sum_\gamma^2$ consisting of those functions which admit a critical point of type $m = y_o^3 + \ldots$ and the corresponding critical value $f(m)$ being equal to another critical value $f(c)$. $\sum_\gamma^2$ does not enter into the Maxwell set, because such a critical point cannot be an absolute minimum, obviously. So $\sum_\alpha^2$, $\sum_\beta^2$ are the only ones occuring in Mx and being of codimension two. $\sum_\beta^2$ is very easy to describe, it is the so-called "triple point". Take three linear functions l_1 , l_2 , l_3 on the plane which are of maximal rank around the origin, and consider the set of points where

$$l_1 < (l_2, l_3),\ l_2 < (l_1,l_3),\ l_3 < (l_1,l_2),$$

and you verify immediately that this gives rise to a triple point when the three linear forms are linearly independent.

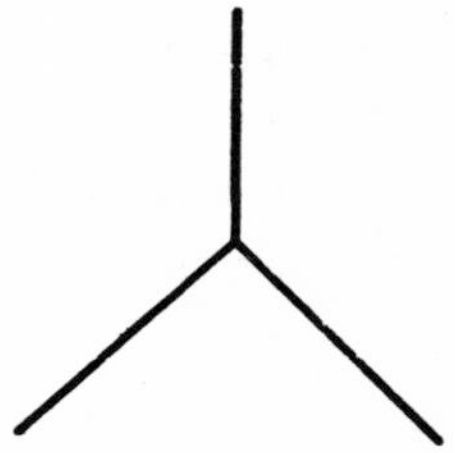

So this is not a very interesting situation, despite the fact that is it very interesting from a morphological point of view: there are many natural phenomena where triple points do occur (e.g. in soap bubbles, in cellular walls in biology, in geomorphology for the summits of a mountain range, etc.), but from a mathematical point of view this is not extremely interesting.

So I will rapidly describe the Riemann-Hugoniot catastrophe. The Riemann-Hugoniot catastrophe occurs when we have the previously described stratum $\sum_{\alpha}^{2}$. So I assume that the absolute minimum of my function is of this type. There are lots of theorems about critical points of real-valued functions which I have no time to explain here, but I will recall a few things very briefly.

First we have to so-called Gromoll's Lemma. It states that if you have a real-valued function $F(x_i)$ defined around the origin of R^n, and suppose it has a critical point at the origin with $F(0) = 0$, then you may write the Taylor expansion as

$$F(x_i) = \underset{\substack{\text{quadratic}\\ \text{term}}}{Q_2(x_i)} + \Gamma(x_j).$$

This quadratic term is in general not a form of maximal rank, otherwise this quadratic point will be non-degenerate. So, in general, the rank of $Q_2(x_i)$ is $n - k$, and I say it is of corank k. This means that I may choose a coordinate system where this quadratic form is

$$x_0^2 + x_1^2 + \ldots + x_s^2 - x_{s+1}^2 - \ldots - x_{n-k}^2$$

(s being the signature of this form). Now Gromoll's Lemma states that I may choose this coordinate chart in such a way that the remaining term Γ depends only on the other remaining variables

$$\Gamma = \Gamma(x_{n-k+1}, \ldots, x_n).$$

Moreover, one can show that, if we consider the restriction of this remaining term Γ to what is obtained by annihilating the first coordinates

$$x_0 = x_1 = x_2 = \ldots = x_k = 0,$$

then one gets what I call in my book the "residual singularity"; one may show that whatever decomposition one chooses, this residual singularity is well-defined up to diffeomorphisms of the subspace, any two expressions of this kind differ only by a change of formulae. (This was not obtained by Gromoll but was obtained by elaborate techniques due to Tougeron and Mather.)

This lemma is very important, because it shows that, in some sense, the quadratic part is really irrelevant; you can always get rid of it by a proper change of co-ordinates. All that is important from the catastrophe theory point of view is the residual singularity; and the number of parameters which occur in it is really the important thing (the corank of the singularity; i.e. the number of relevant parameters which parametrize the catastrophe).

So, coming back to the question of a quadratic point of type Σ^2_α, you see that in this case the residual singularity is y_o^4 (by the Gromoll reduction of the singularity). Now it is very important to understand how the corresponding stratum Σ^2_α maps with respect to the strata of codimension one. In order to understand this, we cut by a two-dimensional plane transversal to this stratum. Now it

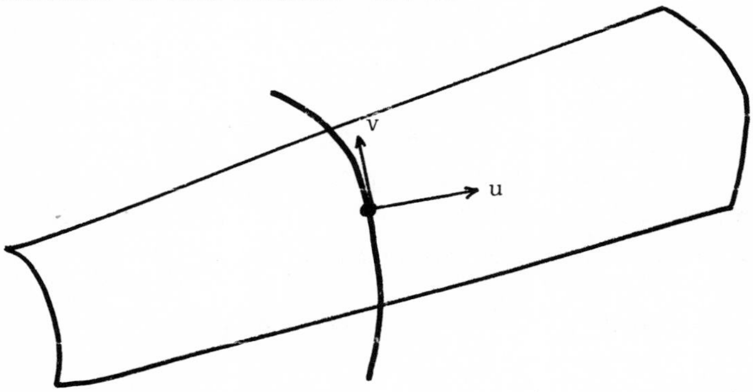

may be shown that, in order to get such a two-dimensional family of functions, transvers to the stratum, it suffices to take the following family

$$\frac{y_o^4}{4} + u\frac{y_o^2}{2} + vy_o$$

(where u,v are local parameters in this transversal section). This whole construction is obviously a local construction, so I can forget about the full structure of the function space, considering only the germ around the origin; this is a purely local affair. I said already that one may show that this two-dimensional family gives some local surface which is transversal to this stratum. This can easily be seen, first that such a section cuts the stratum in only one point, the origin (in no other point would you get a singularity of the same nature), essentially because any function of this kind is obviously a polynomial with three critical points, so that in order to get a singularity of type y^4 it is necessary that the derivative has a triple root (i.e. that $y^3 + uy + v$ has a triple root), which is possible only if u and v vanish. Secondly, one has to prove now that it meets the stratum transversally; since this is a bit less trivial I prefer to omit this here, but it is not difficult to show. (For those of you who know a bit about singularity theory, you know that this family of functions

$$\frac{y_o^4}{4} + u\frac{y_o^2}{2} + vy_o$$

of transverse sections to the strata forms what is called a "universal unfolding" of the singularity. It has the very canonical universal property that any deformation of the germ with an arbitrary large number of parameters may always be mapped onto this transverse section by a family of transformations which are induced by a local change of coordinates. So we get this fundamental result that any deformation of the germ y_o^4 is topologically and differentiably equivalent to the induced family by a mapping of the space of parameters to this universal family.)

This allows us now to give a very simple description of this catastrophe. The derivative $y^3 + uy + v = 0$ admits at most three roots inside the discriminant curve defined by the classical equation $4u^3 + 27v^2 = 0$. So inside the cusp

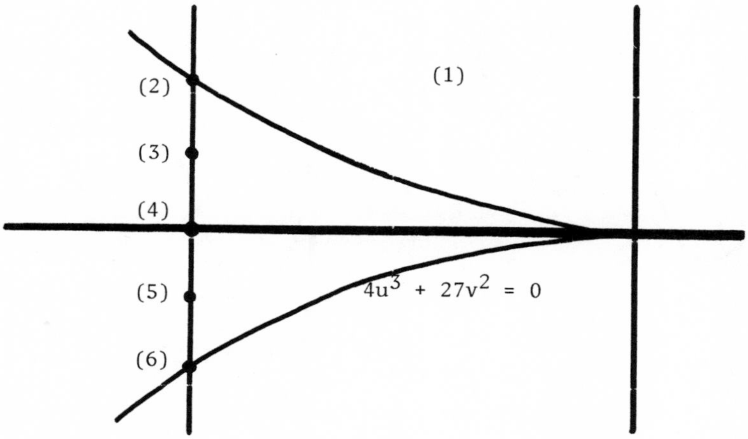

of this curve we have three roots for the equation, and outside we have only one root. This means that outside the corresponding function has only one quadratic minimum (type (1)), and inside the cusp we have the types (2) - (6).

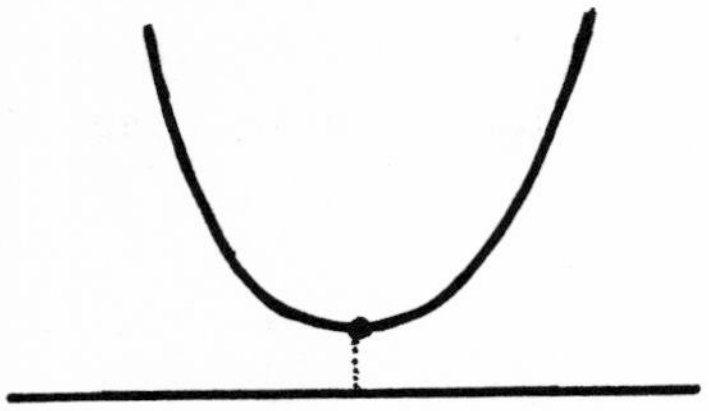

type (1)

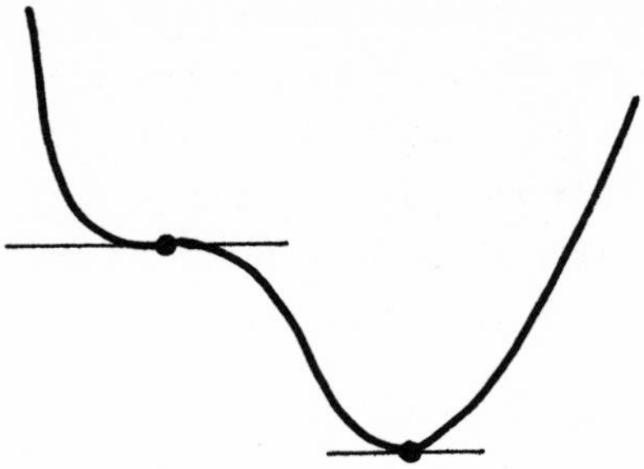

type (2)

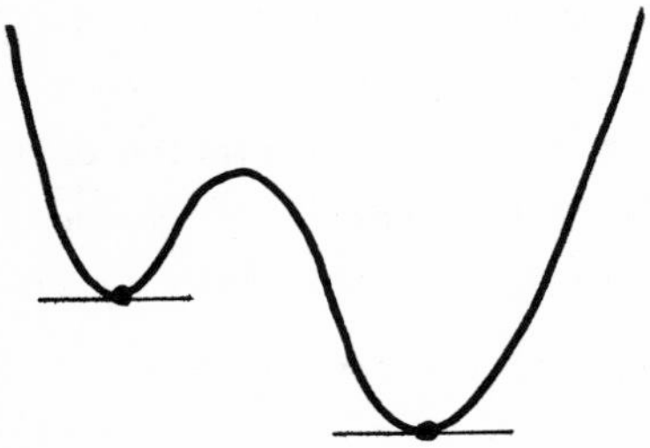

type (3)

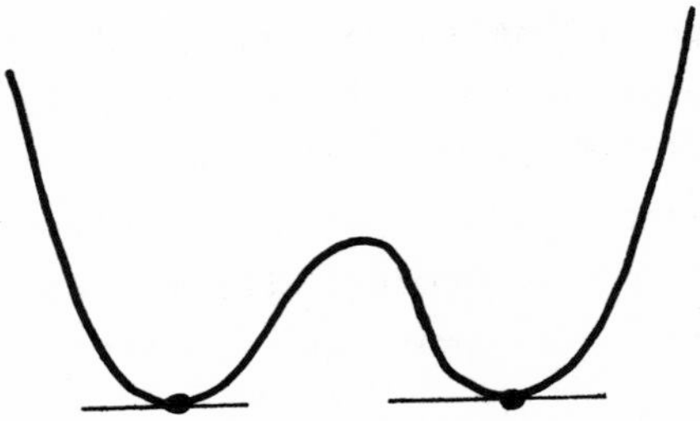

type (4)

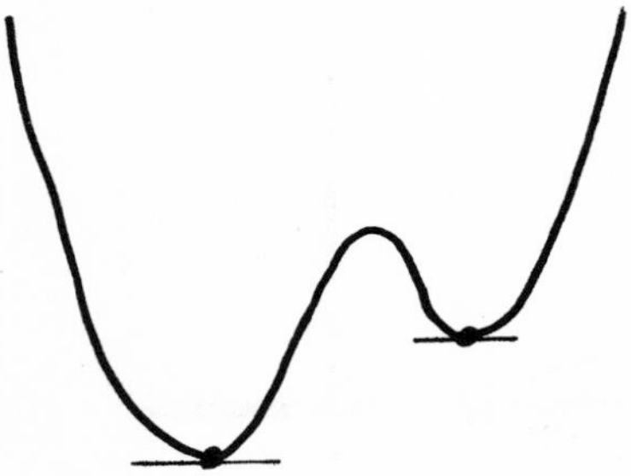

type (5)

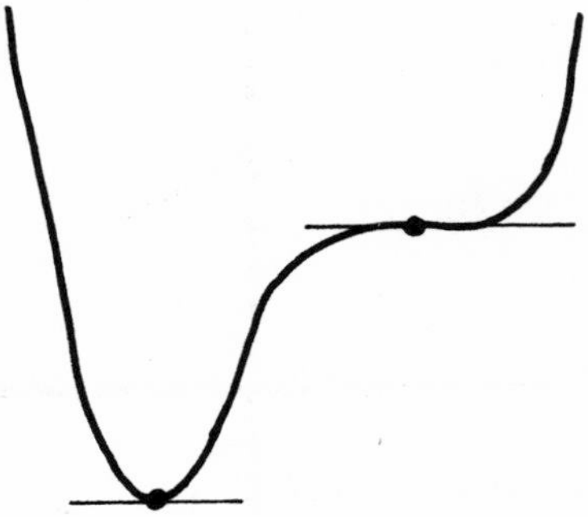

type (6)

Suppose we have a natural process whose local optimization principle is described by such a degenerate minimum at the origin, and suppose further that the local function is given by this equation; what will happen? Suppose I start on the top of the ordinate axis, then I have no choice; there is only one minimum and this point is a regular point of the morphology. While I move down along the straight line

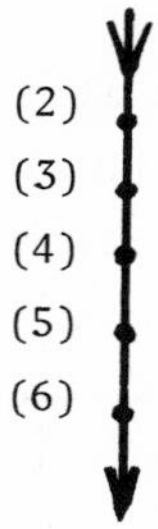

until I reach (2) I am, of course, still at the absolute minimum, and I shall stay in this minimum, not bothering about the appearance of the flex point, thereby reaching position (3). When I reach (4) the new minimum is as small as the old one, so I continue to stay in the old minimum. But if I continue to stay there longer, the new minimum becomes the absolute minimum at (5), until finally in position (6) I will be destroyed and captured by the new minimum. So this shows that we cannot have a continuous regime from (2) until (6), but somewhere in between there has to be a jump from the old to the new minimum. If we follow Maxwell's convention, this jump will occur precisely at (4), along the v-axis, which is the locus of points where the two minima are equal. You see in this example a very typical situation, in which, starting from a completely homogenous situation on the right of the u-axis and moving u to the left, you see some qualitative discontinuity occuring, and a shockwave originating in between two stable regimes. I think that is the basic description of how discontinuous phenomena may be generated by a continuous variation.

This fact is not new. It was already described by Riemann in a paper, I think in 1869, on the propagation of waves, in which Riemann showed that for some quasilinear equation one should not expect its solutions to be continuous on the full plane. But the deeper topological reason for that is the presence of these conflicting regimes, of this catastrophe, and I think that is why it is very important to have a complete classification of all these types of dynamical situations.

Now, if you are interested in processes taking place in usual space-time, then you have to look at strata of the function space which are of at most codimension four, and these form precisely what I have called an "elementary catastrophe"; there are only seven types of them, and the

Riemann-Hugoniot catastrophe is the simplest in the sense explained.

Now I shall say something about the Maxwell convention and the difficulty about using it. One of the big troubles with Maxwell's convention is that it is too mathematical; it gives surface discontinuities of a perfect geometrical clarity. In the case of conflicting regimes in nature, however, you might well have a regime in which a lot of chaotic phenomena take place, and you may not have a shock wave which is a perfectly nice smooth hypersurface. Secondly, you may have a lot of local chaotic turbulence phenomena, and thereby Maxwell's convention may not be valid.

Another reason why Maxwell's convention may not be valid is due to the possible presence of delay. Delay phenomena have to be expected as soon as, in the space of external variables, you have a natural flow, describing the motion of the medium. Suppose the motion is a type of fluid medium, then each part of the medium is a particle of the fluid, and so you may describe a velocity flow describing the motion of this material substrate. Now, if the local regime at one point is a local minimum, then you have to expect this local minimum will stay along the corresponding trajectory of this velocity flow. You also have to expect that this minimum will continue to exist, provided it still is a minimum, even if you have a lower value minimum somewhere else, even if it still exists as a "metastable situation"; i.e. a situation which is locally stable but which is not an absolute minimum. The presence of metastable equilibria is a very natural assumption to make; this actually does occur in many physical phenomema. That is why we said that Maxwell's convention should be replaced in a situation where one has a velocity flow inside the space of external variables. One may use instead what I call the "perfect delay convention", which means that if a local regime has been reached then it stays along the trajectory until the complete breakdown of the corresponding minimum, of the stability of this minimum. This can be seen when the basin of this minimum is described by moving the corresponding potential function. For instance, in the Riemann-Hugoniot catastrophe, if you adopt the perfect delay convention then it may well occur that the velocity flow in the base space is itself a function of the corresponding minimum on the space of internal variables. So, inside the cusp of this, you have two conflicting regimes, and each regime may give rise to a flow. Suppose the dominating regime down has a flow directed upwards and the dominating regime up has a flow directed downwards. What will happen is that, if you move with the perfect delay convention, you will be captured inside the cusp.

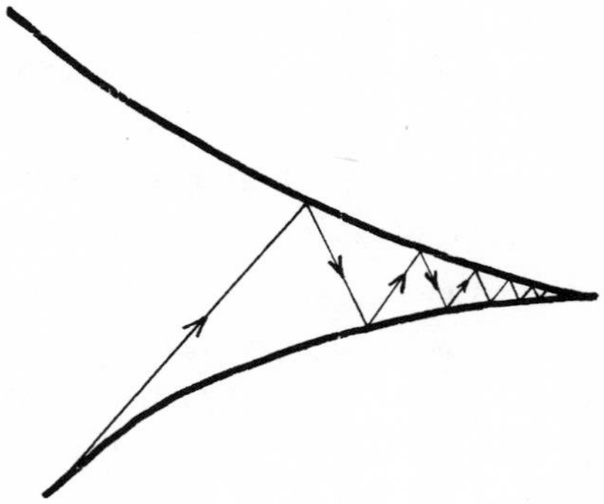

Any material point will be pushed--if you have a little tilt to the right--towards the origin, between the branches of the cusp. So this type of mechanism explains now, how one may get back to the original situation, to the organizing centre of a catastrophe.

I shall explain in my third lecture to-morrow a notion that I call the "hysteresis cycle". This is again a very natural notion which often occurs in nature, and I may describe such a hystersis cycle in the following way. Consider the curve $y^3 + uy + v$ and suppose we take for u some negative value, say $u = -1$. This curve has the following appearance:

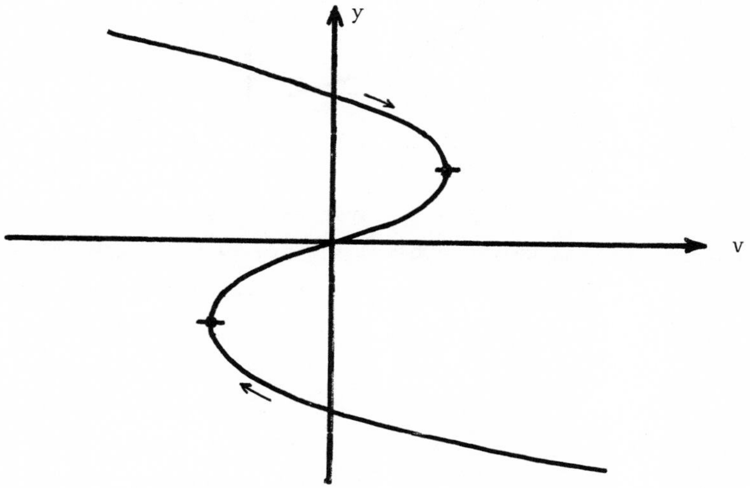

The upper and lower branches describe, as usual, minima, so these describe stable regimes, and the intermediate branch describes a maximum; i.e. an unstable regime, which does not exist. Now a hysteresis cycle is obtained as follows: Suppose that from the lower minimum branch you have a flow which moves the representative point to the left, and on the upper branch you have a flow which moves the representative point to the right. Then, if you start at the lower branch, you will be moved along this stable branch until you reach this critical point, where you have no longer the possibility of staying on the branch. You will be pushed by a massive jump to the upper existing minimum, and here again you will be moved by the flow to the other critical point and pushed back down to the lower minimum, and this completes the cycle. If u is fixed you see that a representative point oscillates indefinitely on this segment in the universal unfolding space.

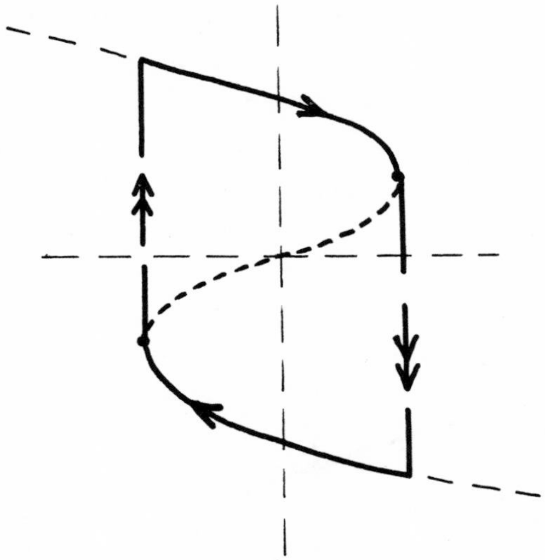

I shall describe to-morrow how this hysteresis cycle may be generated by a very natural process which involves the presence of hyperbolic metrics, and I will give-- if I have time-- some applications to biology.

Third Lecture

Yesterday I introduced the notion of hysteresis cycle. Such hysteresis cycles can be generated by a fairly natural type of process, and I believe that this type of process has some philosophical interest. Up to now people in mechanics have always believed that when you want to study the evolution of a system you fix the configuration space or the phase space and you vary the potential. If something changes in the evolution of the system then it is assumed that the configuration space has to be fixed, that the potential function or the Hamiltonian function may vary with time. The model I will describe to you is inspired, I would say, by another kind of philosophy, namely it is the idea that the potential function might well be constant but that in some sense the space changes (by space I mean essentially the underlying metric).

Let me come back to the Riemann-Hugoniot catastrophe whose potential function was given by that polynomial of degree four:

$$V = \frac{y^4}{4} + u\frac{y^2}{2} + vy.$$

Recall that if I write down the derivative with respect to y

$$\frac{\partial V}{\partial y} = y^3 + uy + v$$

and if I put $u = -1$, in order to get real roots for the derivative, then I get the classical wriggle.

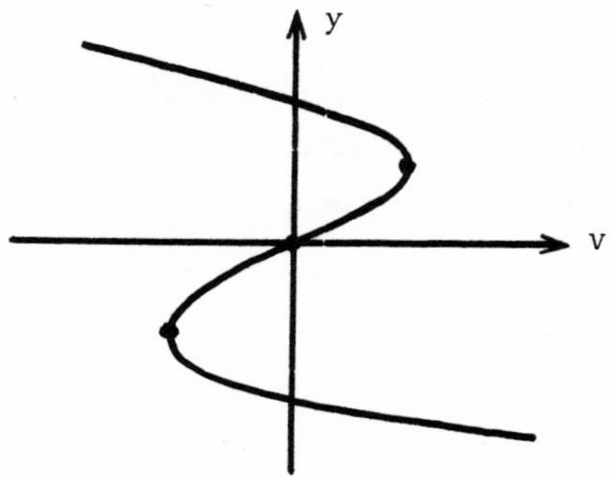

How can a hysteresis cycle be generated by not too unnatural assumptions? The basic assumption one can make is as follows. Suppose you start with a local potential which is of the type having only $\frac{y^4}{4}$. This is a degenerate potential, so this unstable situation will be destroyed, and the problem here is to understand how such an unstable situation can be destroyed. Here we meet with a very general type of question, namely about the problem of "first integrals" in dynamics. You know that, if you have a fibration, say, from the manifold M^n to the base manifold B, then a flow, a vector field X, will be said to be "vertical" if it is tangent to the fibre $r^{-1}(b)$. The existence of a first integral for a dynamical system amounts always to saying that--at least locally-- the phase space admits a local fibration (or a local foliation) which is kept invariant by the flow; the flow is tangent to the leaves of this foliation. There is a very general result (which has been proved by a student at IMPA, I think his name is Manè) that, as a corollary of a result of Takens, one can show that, for any compact manifold, if we form the space $X(M^n)$ of all flows, then the set of those vector fields which admit a smooth first integral, even with singularities, is nowhere dense. In other words, the property of admitting a first integral is not a stable property of a dynamical system. Of course, this raises very interesting questions from a philosophical point of view, because in physics the notion of first integral plays a very important role. So, if one believes that genericity assumptions are valid physically the notion of first integral should not exist. Why is it that we get first integrals in mechanics? This is a very far-reaching problem which deals basically with the question of understanding the meaning of Hamiltonian dynamics in nature in general. It is believed that all physical laws are expressed by Hamiltonian (i.e. time-reversible) dynamics, contrary to the obvious fact that time is irreversible. You know that the only way of getting out of this difficulty has been to--how do you say in English?--to push the dust under the rug, by saying that the initial data were extremely singular, that the origin of the universe was a huge "big bang", that there was a huge catastrophe at that time and that we are still running on the effects of this original catastrophe.

I believe myself that this is a very poor answer to this fundamental problem of understanding why physical laws are time-reversible despite the fact that time is obviously so irreversible. But, of course, first integrals may be shown to be perfectly stable in the sense of statistical stability in Hamiltonian dynamics. This is why--in the framework of Hamiltonian dynamics--the presence of first integrals is not so astonishing. I may refer also to the so-called Noether theorems which state that if you have a Hamiltonian system which is invariant under Lie group action then the dual of the Lie algebra of this group forms at least a system of local first integrals for this dynamical system. As soon as one gets symmetries in nature one also gets first integrals, of course. And also the problem of the existence of symmetries in nature is obviously one of those questions which cannot easily be dismissed. There are many local phenomena which exhibit symmetries. So here we deal with one of the really basic mysteries of nature, but mathematically there is absolutely no mystery at all. Takens proved that in the space of all vector fields on the manifold there is a Baire subset which has the property that the Ω-set (i.e. the set of non-wandering points of the flow) is a continuous function with the Hausdorff topology of the flow itself. So, on the Baire set of the function space the set of non-wandering points is a continuous function of the flow, and as a result you may show easily that, if you have a smooth first integral, then take a regular value of this first integral, this gives you a local fibration which is kept invariant by the flow. Then, by a little perturbation transverse to this foliation, you get a big destruction of the Ω-set, because originally the Ω-set had to meet each one of the fibres; but, as soon as you apply this local perturbation, all the points of the fibres start wandering, and you destroy the Ω-set to a large extent. This is just a sketch of the idea of the proof. If the first integral would be stable then the Ω-set had to persist, that way getting a contradiction.

So this shows that, in general, first integrals are not stable in nature and here, in our catastrophe theory framework, we deal with a situation in which we have initially the space of external variables, the space of internal variables, and the basic dynamics originally is a gradient dynamics inside the space of internal variables. This admits as first integrals the fibration defined by the projection onto the space of external variables. Obviously this is an unstable situation, so one has to ask how such an unstable situation might be reasonably destroyed; what is the most natural way of destroying a first integral? The answer to that, I think, can be given as follows. We have to accept the idea that unfolding a catastrophe is in some sense a way of damping the catastrophe. We have a sort of action-reaction principle which has the effect that unfolding the original situation tends to damp the effect of the original dynamics. If we make this type of assumption we shall say that we have a vertical flow at the origin, then there develops a horizontal flow inside the product space. The most natural horizontal flow to develop is one which will damp the effect of the original dynamics. If we write down this condition we get that the scalar product $\langle H,\xi\rangle$ of the horizontal component with any vector should be

$$\langle H,\xi\rangle = -\lambda^2 \langle \text{grad } V,\xi\rangle \quad .$$

Accepting this definition for the horizontal component one has to take the gradient of V with respect to a hyperbolic metric $dy^2 - \lambda^2 dv^2$. If you look at what happens around the origin you find that you may forget about the higher order term around the origin, you just have something like $V_1 = vy$. If you compute the gradient of V with respect to the hyperbolic metric $dy^2 - dv^2$, then you will find easily that this vector has as components V and $-y$, so this is nothing but a flow defining a rotation around the origin. This is an interesting fact; so you see that in some sense such a flow can be considered as a gradient with respect to a hyperbolic metric. I am wondering whether in many situations in physics, when people believe they deal with Hamiltonian situations, they are not in fact dealing with gradients with respect to hyperbolic metrics. This is a problem on which I have no precise idea for the time being, but certainly it would be worth investigating whether or not there is a sort of equivalence between these two viewpoints.

I myself believe that this point of view of hyperbolic metrics is perhaps more interesting, because of the fact that, if we now consider the remaining term as a perturbation of this quadratic term vy, then we see that the trajectories of the gradient of vy are the circles around the origin. If you now add the extra term $y - y^3$ tou see that, if we restrict ourselves to positive y, you get for small y in this halfplane $y > 0$ a positive term, and for large y you get a negative term. Similarly, in the halfplane $y < 0$, you get a negative term for small y and a positive term for large y.

This means that if you move along the trajectory you do not come back to the original situation, but you have a vertical displacement. This shows that, in between, there exists a point on the y-axis where the total shift is zero; such a point will, of course, give rise to a closed trajectory of the flow.

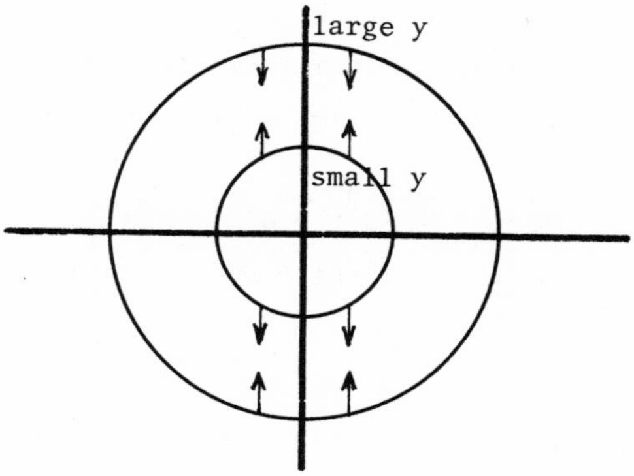

This closed trajectory will be an attracting trajectory, because both inside and outside you are moved toward it. This shows that this gradient flow has a limit cycle which cuts the y-axis approximately at the point where y and y^3 are of the same magnitude. This is, of course, not a complete proof of this fact; this has been considered in the theory of the van der Pol equation, and I have been told by specialists (e.g. by Arnold in Moscow) that proving rigorously that you have only one closed trajectory requires a lot of computation and the use of special functions, like elliptic functions. Anyway, for us this very qualitative argument is sufficient, and so we arrived at the result that this gradient flow with respect to a hyperbolic metric has a closed trajectory which is a stable limit cycle.

Now suppose that we continue the degeneracy of our metric by letting λ go to ∞. I want to normalize the metric by using the factor $1 - \lambda^2$, and let λ tend to infinity. The underlying philosophy of this type of consideration is that initially you had a product space on which only the y-coordinate was relevant for the study of the dynamics, and at the end of the catastrophe you have only the v-coordinate relevant for defining the dynamics of the system. So it describes, so to speak, the death and birth of interesting parameters, and I believe that this is the kind of consideration which has never been introduced systematically in the study of natural phenomena. Quite likely, this type of consideration may be used for the study of many evolutions of natural processes. The idea that some co-ordinate might be relevant at some time and become irrelevant at some other time is a very natural idea, and I think it certainly may be used to explain many types of phenomena.

Now, what does happen when λ tends to infinity? One sees almost without any difficulty that then one has either to suppose that

$$\frac{\partial V}{\partial y} = 0$$

and then the components in y are arbitrary, or one has to assume that

$$\frac{\partial V}{\partial y} \neq 0$$

and then the component in y is 0. This shows that, when we tend towards the degenerate metric, the only possibility is that either the representative point stays on the branch of the curve $\frac{\partial V}{\partial y} = 0$, or, if it is outside this stable branch, then the only thing it can afford is to be vertical. This shows that this limit cycle, when the metric degenerates, tends toward the hysteresis cycle which I described in the last lecture; so the hysteresis cycle can be obtained by a limit process of a gradient with respect to a hyperbolic metric, letting this metric degenerate.

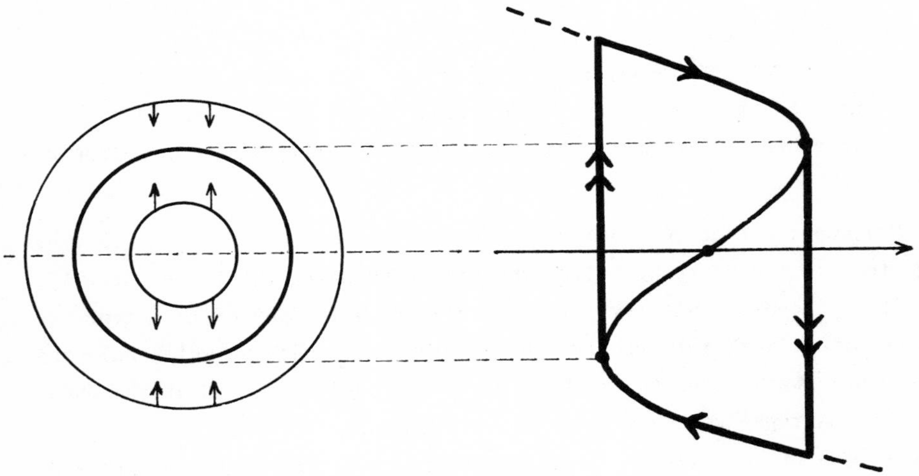

If you allow me some speculation, I think this kind of consideration might explain the importance of hyperbolic metrics in physics. In relativity, for example, when one deals with the metric

$$c^2dt^2 - dx^2$$

(t time, x spatial length, c velocity of light), I believe that this might express the fact that it is perhaps naive to believe that, if you consider space-time locally as a product of space and time, then the evolution of the universe, of a locus of simultaneity, should be like the evolution of a nice plane wave front along the time axis. It is perhaps more natural to believe that this evolution is not so regular, and that almost periodically you get some phenomena, that the locus of simultaneity may get wriggles of this type, so that at some times time might be a non-uniform function. If you accept this kind of idea you will admit that sometimes you get a singularity with respect to time so the singularity has to be unfolded by space. So one will get a metric involving the square of time minus the square of spatial space in this process. If we have a metric only on space, we will get vertical trajectories at the limit, meaning that we will get punctualization; i.e. localization of objects. Conversely, one would get trajectories involving instantaneous propagation. I cannot believe that this to some extent might explain the fundamental duality one meets in physics between matter and radiation. Matter is something which really can be localized and radiation is something which thinks only of flowing away. Of course, in quantum physics people say that one has a nice formalism to put the two things in the same mathematical framework; but that is also a nice way of pushing the dust under the rug. I believe that the distinction between matter and radiation is something fundamental, it is a distinction of a qualitative nature and one really has to give an explanation for it. Well, so much about the hyperbolic metric.

I have presented this study very roughly, and only for the case of the Riemann-Hugoniot catastrophe-- it will be interesting to see what happens for more complicated catastrophes, and what kind of object generalizing these hysteresis cycles are obtained when one considers more complicated singularities.

Now I will perhaps give some applications of this. This is, of course, very speculative; but it may have much relevance for the study of embryology, and biological regulations. I am thinking about a basic fact which occurs in embryology. Consider, for instance, the egg of the frog. At the beginning this is one cell, then this cell starts dividing, and after some time it gives rise to a kind of hollow sphere, called the blastula. Why is it that it develops a hole inside? This is something that is still very mysterious, and biologists have no proper explanation for it. Of course, this hole is not empty, it contains some juice, some kind of liquid, but nevertheless the cells form a kind of sphere with a thickening in the lower part of the blastula; this thickening contains the reserves which are necessary to construct the embryo; that is the so-called yolk.

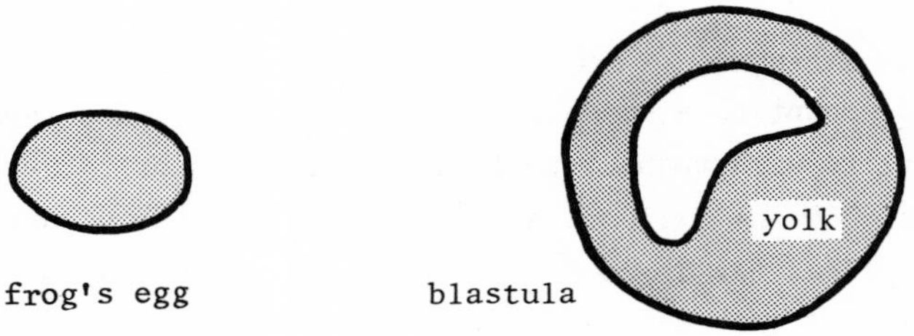

Now, after some time, a very abrupt morphological accident does occur which is known as the phenomenon of gastrulation. It can basically be described as follows. At some time a kind of furrow is formed, and a full layer of external peripheral tissue goes through the egg; it is an invagination as they say in biology. When you look at this from the dorsal part of the egg you see first a kind of furrow in the form of a crescent and after some time this crescent develops and finally closes itself.

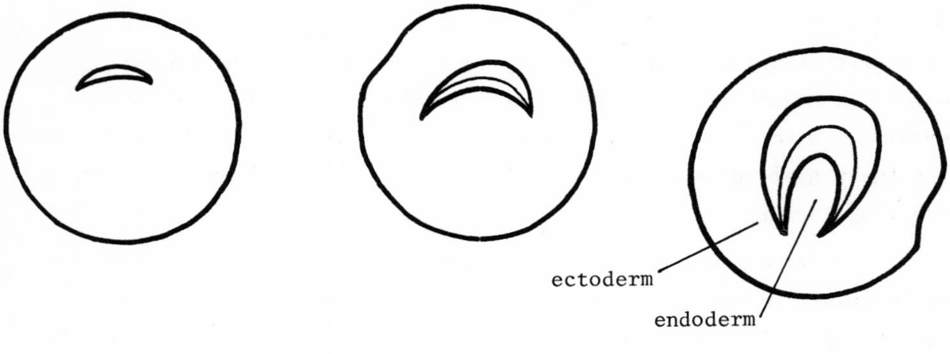

On the bottom of this crescent you have a kind of shock wave separating two kinds of tissue, and when this crescent is closed these shock waves are separated completely; externally you have the so-called ectoderm and internally the endoderm. Moreover, another type of phenomenon occurs, namely cells are moving and getting back inside, so that after some time the situation looks practically as follows. Schematically speaking, the egg becomes a tri-layer structure.

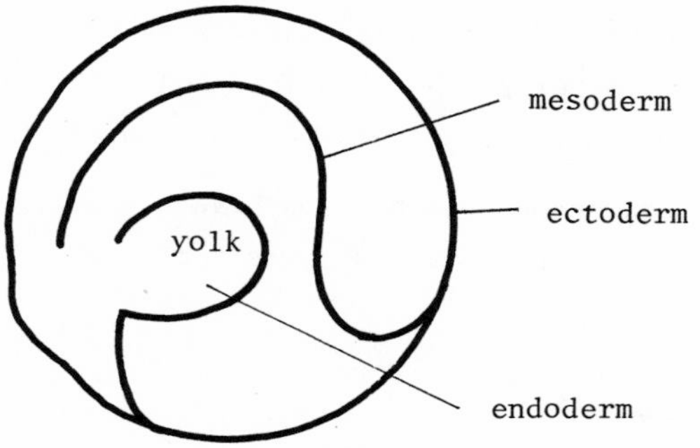

This decomposition of the tissue into three main layers (ectoderm, mesoderm, endoderm) is a very general rule in animals, in particular in vertebrates, so that biologists say that the embryology of all these animals is triploplastic (i.e. you have a three-layer embryo after some time). The only exception to that are very primitive animals, such as the hydra, which are only diploplastic and do not have mesoderm. It is quite interesting to see what these three layers give in developed state; to save some time I will not entirely describe the full process. But, basically, endoderm gives very little except the intestinal mucosa (i.e. the inner lining of the digestive tract, that part which assimilates the food). Ectoderm gives essentially the skin and the neural system (nerves, brain, spinal chord), and also the sensory organs. Mesoderm gives the bones and the muscles. Of course, a part of mesoderm also gives skin (derm), whereas epiderm comes from ectoderm. I shall not describe this in more detail.

What is the meaning of this general decomposition of an embryo into three layers? I think that I could give the following general scheme of explanation. The basic requirement for regulation, for homeostasis of an animal is feeding; you have to restore your reserves and energy, so you have to get some food. For most animals, getting food means capturing a prey. The restoration occurs by an abrupt catastrophic process, namely capturing a prey. This is a big difference between animals and plants. In plants, restoration of energy occurs by the use of solar energy, the chlorophyl being used in order to capture chemical energy; so that is a fairly continuous process, except that of course it can happen only by daytime and not during night. But for animals the problem is much more complex, because the animal has to capture a prey, and this is more complicated because the prey is another object, essentially another animal or a plant, in any case another living being. So this fundamental requirement here shows that a basic morphology for animals is the "capture morphology".

If you draw a global scheme of this you have a predator and the prey, and what happens is that the predator captures the prey.

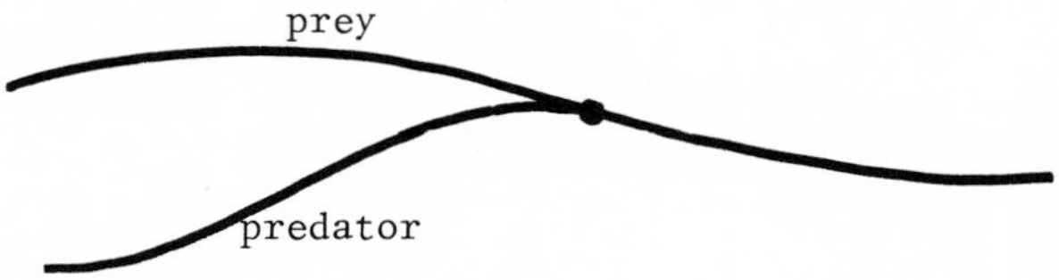

In general the predator will have to move in order to capture the prey, because the prey generally tries to escape the predator; so the global morphology of feeding is the capture morphology. In order to describe it, if we look at it as a catastrophic process, the simplest type of singularity which described the capture morphology is the Riemann-Hugoniot catastrophe. In this catastrophe we had a fourth degree polynomial, and we can change the parameters of this polynomial in such a way that the basin of the small minimum is captured by the lowest minimum. In order to get this kind of evolution I have to consider a straight line intersecting the corresponding discriminant curve; at the intersection point K the capture takes place.

One sees immediately that such a process is fundamentally irreversible, so, as a new prey is met one has to be able to get back to the original situation. So, if you want to look at feeding as a periodic activity, you have to close this irreversible process.

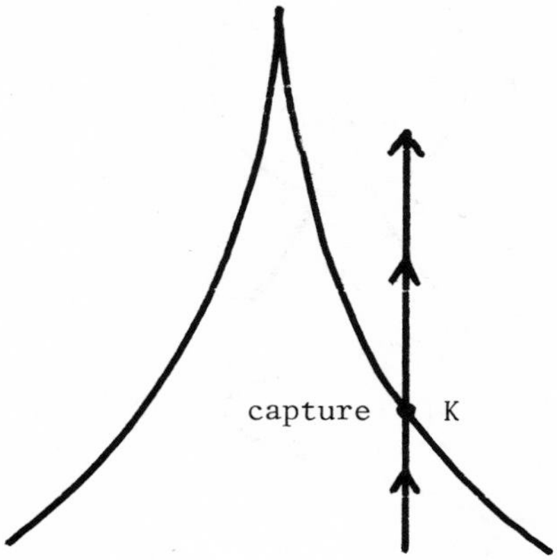

The simplest way of doing this is by replacing the local straight line by a closed loop around the origin as follows.

At the point K I have the capture morphology as before, but you see that something very strange occurs now: If I move along the closed loop and come back to the point J then, by moving around the critical point we have exchanged the stable branches above, so that at J the predator comes back as the prey; a new kind of catastrophe takes place at J, namely, one gets a new branch. If I draw the corresponding counterimage in the space of internal variables I get the following:

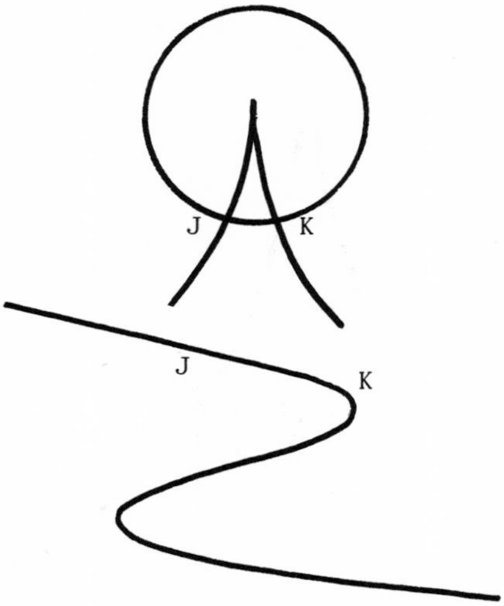

This model shows that, after turning around the origin, the predator (who was satisfied before, because he had sufficient energy after he had fed) becomes hungry again after some time, when his reserves of energy are getting down. And when he becomes hungry then--according to this scheme--he becomes his own prey. This is a very strange idea, at first glance; how is it that a predator becomes his own prey?

When I first saw this requirement I was quite struck, and I thought this was completely silly and I should give up this idea. But after thinking it over I found that, on the contrary, this is a highly interesting idea. Namely, it shows that the predator has to recognize his prey among all the objects which are in the neighbourhood, and I think it is basically a fundamental requirement that you cannot recognize anything but yourself. By that I mean that any kind of recognition process requires the use of some resonance phenomenon; you have to produce a resonance between an external form and some internal form which lies already in your brain. Any resonance phenomenon requires some metric equality between the period of the impinging stimulus and the period of the internal oscillator. So, in that respect, I believe that one cannot recognize anything but oneself; and so one has to look at what happens at point J as a very new kind of catastrophe that I call the "perception catastrophe": as soon as the predator recognizes an external prey he drops down on the lower level (in the preceding figure) and starts moving in order to capture the prey; he becomes himself. Before he was his prey, but as he has recognized an external prey he becomes himself and starts moving to capture the prey.

Those among you who have some philosophical background may read a book by Husserl entitled "Erfahrung und Urteil". You will see that in this book Husserl described almost explicitly this kind of process. He says that recognizing the object is nothing but grasping it. In this process of grasping, the ego, the subject, becomes himself.

You may, of course, not accept this kind of strange idea, but I believe that this sort of "consubstantiation" between predator and prey has a very deep importance in many fields. First, in biology itself, because there are many instances in which the predator has a kind of simulation of many properties of the prey. (Consider, for instance, an animal such as the lamprey, having a tongue with an extremity of the shape of a small worm; by moving this wormlike extremity of the tongue it attracts a small fish around it and then captures its prey by a big ingestion of water). There are many instances of biological morphologies, or organogenesis, where you have this simulation of the metabolism of the prey by the predator, and this is something really very difficult to explain by the standard explanations which biologists are used to give, such as Darwinism, or Neo-Darwinism. I believe that the probability of an animal developing something that looks like the prey's prey is highly unbelievable, unless there is an underlying mechanism to explain it.

Moreover, this kind of consubstantiation between predator and prey might explain also the almost universal belief among the so-called primitive populations that man may transform himself, if he has some sorcery gifts, into animals. That is a belief still accepted in many countries of the world, and it is very easily seen that the animals for which this transformation occurs are those which are in relation of predation to man; either they are preys (animals that man hunts), or they are predators of man (such as tigers or sharks). If you look at the ethnological literature you will see that this transformation of man into animals occurs very easily precisely for those animals which have a very strong predation relation to man. In many tribes, for instance, when the men start hunting some kind of animal, they will start to hunt by a ritual which almost always involves simulation of the prey (e.g. by putting on the skins of such animals, or by jumping like the animals). It seems to be a very deeply ingrained belief that, in order to capture the prey, one has to become like the prey oneself. Well, of course, spatially it is not possible to become one's own prey, so there is an organ which, in some sense, smoothes this catastrophe of being something else than oneself. This organ is nothing but the brain. The central nervous system is the organ which allows an animal to be something else than himself; it is an alienation-permitting organ.

But let me come back now to the problem of mesoderm. This is a very interesting problem, because when one looks at the formation of the intermediate layer, that wriggle I was describing, one would expect the mesoderm to be unstable, but in fact

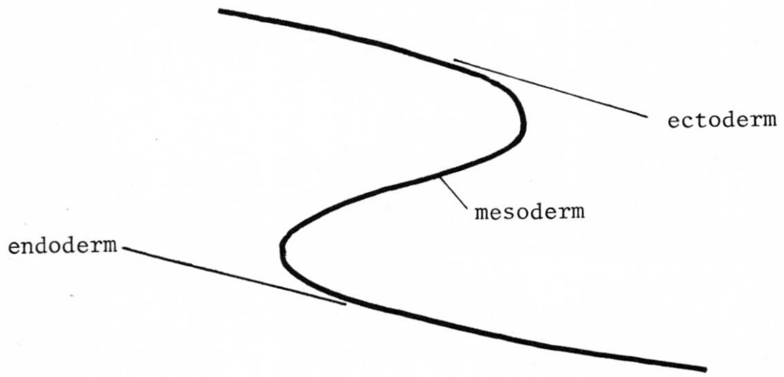

after some time it becomes stable. But it becomes stable only locally, and it assumes very specific positions and shapes. The first position of stabilization of mesoderm is the central meridian, the plane of symmetry of the animal, where an organ is formed which is called notochord; one could say it is the sketch of the skeleton. My point is that this stabilization of this intermediate layer occurs precisely through this formation of a limit cycle which I described earlier. Basically, the distinction between ectoderm and endoderm arises through a Riemann-Hugoniot catastrophe, but mesoderm is stabilized by the formation of a limit cycle, and this limit cycle is transformed towards a hysteresis cycle which has a very obvious interpretation. If you look at this complete "wriggle" as the content of energy of the animal (locally), then the corresponding hysteresis cycle can be interpreted as follows: the animal uses his own reserves of energy in order to push the prey towards the capture point; so the process of capturing the prey requires some expenses on the energetic reserves. This is precisely done by the hysteresis cycle: you loose energy on the lower level in order to bring the prey towards the capture point.

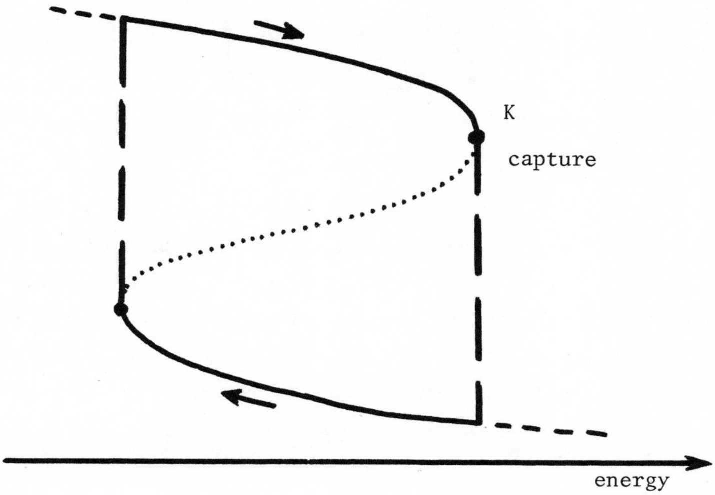

Well, I have not enough time left to explain this in more detail; one can push this model very far. Initially this notochord is forming a metabolic cycle which has the effect to induce on the ectoderm the formation of the nervous system which is this smoothing organ for the perception catastrophe. When one looks at this situation, one has a conflict between two gradients, one descending gradient for the nervous system and one ascending gradient for the mesoderm; so in between one gets the formation of a lot of little convection cells (see figure on the next page). This is a model for describing the periodic formation of vertebrae along the vertebrate axis. One gets this kind of local cycle on the nervous system, and this cycle describes precisely the evolution of the information inside the nervous system. The nervous system gets information from the sensory organs and then sends back this information towards the motor system of mesoderm (muscels, bones) in order to capture the prey. So this little cycle is more or less a part of the huge functional cycle of evolution, of transformation of information of the sensory organs towards the motor organs.

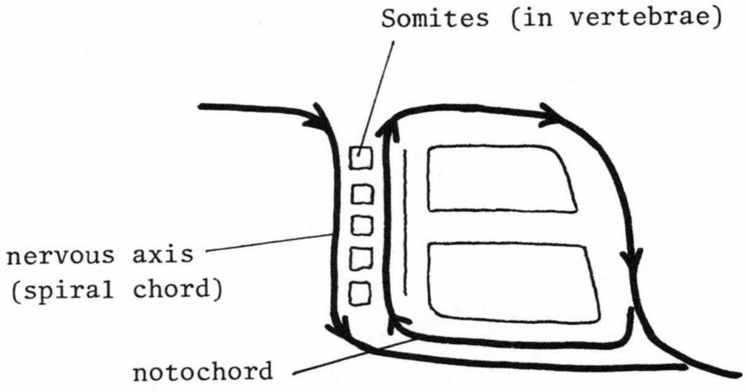

And here we also have the metabolic implication; the kidneys (a part of mesoderm), being organs which realize a sort of filtering process of the blood, are really bifurcating organs. A bit further down on the mediolateral gradient one finds a piece of mesoderm which in some sense realizes organically this cycle; it consists of two sheets, an upper part called the somatopleur, and a lower part called splanchnopleur;

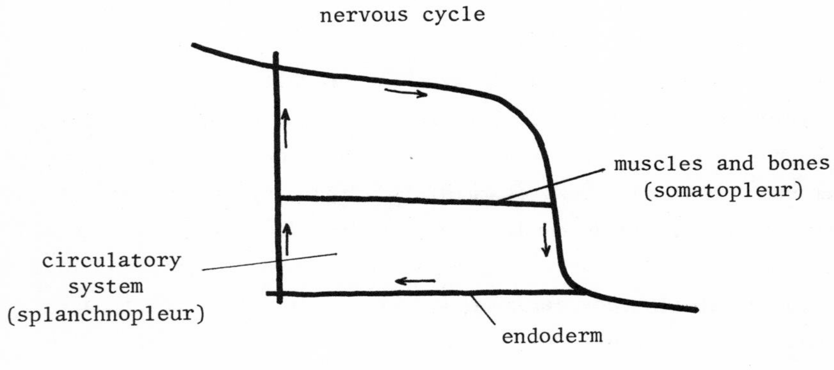

the upper part will form bones, muscles and limbs, and the lower part will form blood and the vascular system. So the blood is more or less the part which is moved by this full cycle, because extracting energy requires motion of the liquid. Basically one has two kinds of cycles: the information cycle in the nervous system, and the blood circulation, which is circulation of oxygen and energy, on the other part.

Well, I think I have said enough just to show you what kind of ideas one may get when one applies topology to biology.

BIBLIOGRAPHY

The items listed here should be of use to those who wish to follow the theory of catastrophe further:

1. R. Courant and K.O. Friedrichs: Supersonic flow and shock waves, Interscience 1948.

2. D. Fowler: The Riemann-Hugoniot catastrophe and Van der Waals' equation, C.H. Waddington (ed.), Towards a theoretical biology, vol. 4, Edinburgh Univ. Press 1972.

3. A.N. Godwin: Three dimensional pictures for Thom's parabolic umbilic, Publ. I.H.E.S. 40 (1972).

4. A.L. Hodgkin and A.F. Huxley: A quantitative description of membrane current and its application to conduction and excitation in nerves, J.Physiol. 117 (1952).

5. A.L. Hodgkin: The conduction of the nervous impulse, Liverpool Univ. Press 1964.

6. J.N. Mather: Stability of C^∞ mappings I, Ann. Math. 87 (1968).

7. - : Stability of C^∞ mappings II, Ann. Math. 89 (1969).

8. - : Stability of C^∞ mappings III, Publ. I.H.E.S. 35 (1969).

9. - : Stability of C^∞ mappings IV, Publ. I.H.E.S. 37 (1970).

10. - : Stability of C^∞ mappings V, Advances in Math. 4 (1970).

11. - : Right Equivalence, (mimeographed), Univ. of Warwick 1972.

12. T. Poston: The Plateau problem, Summer College on global analysis and its applications, Internat. Centre for Theoretical Physics, Trieste 1972.

13. T. Poston and A.E.R. Woodcock: Zeeman's catastrophe machine, Proc. Cambridge Philos. Soc. (to appear).

14. - : The cuspoids, Springer Lecture Notes (to appear).

15. - : The elliptic and hyperbolic umbilics, Springer Lecture Notes (to appear).

16. T. Poston and A.E.R. Woodcock: The parabolic umbilic, Springer Lecture Notes (to appear).

17. R. Thom: Stabilité structurelle et morphogénèse, Benjamin 1972.

18. - : La stabilité topologique des applications polynomiales, Enseignement Math. 8 (1962).

19. - : Ensembles et morphismes stratifiés, Bull. Amer. Math. Soc. 2 (1969).

20. - : Topological models in biology, Topology 8 (1969).

21. - : Topologie et linguistique, Haefliger and Narasimham (eds) de Rham commemorative volume, Springer 1970.

22. - : A global dynamical scheme for vertebrate embryology, Amer. Math. Soc. Audio Recordings, no. 70 (1971).

23. D'Arcy Thompson: On growth and form, Cambr. Univ. Press 1961.

24. C.H. Waddington: Principles of embryology, Allen and Unwin 1956.

25. C.T.C. Wall: (ed.) Proceedings of the Liverpool Singularities Symposium I, Springer Lecture Notes 192 (1971).

26. - : (ed.) Proceedings of the Liverpool Singularities Symposium II, Springer Lecture Notes 209 (1971).

27. H. Whitney: Local differential properties of analytical sets, Differential and combinatorial topology, Princeton 1965.

28. A.E.R. Woodcock: Morphogenesis and catastrophe theory (to appear).

29. A.E.R. Woodcock, T. Poston, and I.N. Stewart: The Double Cusp, Springer Lecture Notes (to appear).

30. E.C. Zeeman: Breaking of waves, D. Chillingworth (ed.) Symposium on differential equations and dynamical systems, Springer Lecture Notes 206 (1969).

31. - : Differential equations for the heartbeat and the nervous impulse, C.H. Waddington (ed.) Towards a theoretical biology vol. 4, Edinburg 1972.

32. - : A catastrophe machine, C.H. Waddington (ed.) Towards a theoretical biology vol. 4, Edinburgh 1972.

33. - : The geometry of catastrophes, Times Lit. Supp. December 1971.

LOCALIZATION OF NILPOTENT SPACES

Peter Hilton*

Battelle Seattle Research Center

1. Introduction

The technique of localization was first introduced into topology by Sullivan [11], though it was implicit in Zabrodsky's method of mixing homotopy types [12]. Subsequently it has been exploited by many topologists, e.g., [1,5,7,8,10]. The author, Mislin and Roitberg [5] have used the technique extensively in studying non-cancellation phenomena. A comprehensive treatment of a more general process, executed in the semi-simplicial category, is given in [1]. We give two examples to show the potential of the method.

Example 1.1 Let $V = V_{n+1,2}$ be the Stiefel manifold of unit tangent vectors to S^n. Then V fibres over S^n with fibre S^{n-1} and it follows from a classical theorem of James and Whitehead that V admits a cellular decomposition.

$$V = S^{n-1} \cup e^n \cup e^{2n-1}. \tag{1.1}$$

Moreover, if n is even, then the first attaching map in (1.1) has degree 2 (the Euler characteristic of S^n). Now, as will transpire in Section 2, we may localize cellularly. Thus if we localize at the odd primes, P, then $S^{n-1} \cup e^n$, in (1.1), becomes contractible (since 2 is invertible in the ring of integers localized at P), so that we obtain, from (1.1),

---

* This is a report on joint work with Guido Mislin and Joseph Roitberg. An expanded version, under joint authorship, will appear as a monograph in the series *Notas de Matematica*.

$$(1.2) \qquad V_P \simeq S_P^{2n-1}.$$

Of course, the implications of (1.2) for the cohomology of V were already well known. However, (1.2) also enables us to conclude that, for any space Y which can be P-localized, the set of homotopy classes of maps of V into Y_P satisfies

$$(1.3) \qquad [V,Y_P] \cong \pi_{2n-1}(Y)_P.$$

Thus, in particular, $[V,Y_P]$ has an abelian group structure. In general, of course, $[X,Y]$ is merely a set with distinguished element, and thus very difficult to handle.

One may say that the traditional tactic in algebraic topology has been to apply an algebraic functor (e.g., homology, cohomology) and then localize at some prime. By localizing first, we may gain structure, as in this example.

Example 1.2 Let $S^3 \rightarrowtail E \to S^7$ represent a principal S^3-bundle over S^7. Such a bundle is classified by an element $\alpha \in \pi_6(S^3)$. Now $\pi_6(S^3) = \mathbb{Z}/12$, generated by ω, the Blakers-Massey element which expresses the non-commutativity of the group operation (quaternionic multiplication) on S^3. We will write E_k for the total space E of the bundle classified by $\alpha = k\omega$, $0 \le k \le 11$. Of course E_k is diffeomorphic to E_ℓ if $k + \ell \equiv 0$ mod 12. However, we may prove, by a cellular approximation argument, that $E_k \not\simeq E_\ell$ unless $k \equiv \pm\ell$ mod 12. For, by the James-Whitehead theorem cited above, E_k admits a cellular decomposition

$$E_k = S^3 \cup_{k\omega} e^7 \cup e^{10}.$$

Thus if $E_k \simeq E_\ell$ then $S^3 \cup_{k\omega} e^7 \simeq S^3 \cup_{\ell\omega} e^7$, and from this we rapidly deduce (using a classical desuspension theorem) a commutative square

$$\begin{array}{ccc} S^6 & \xrightarrow{k\omega} & S^3 \\ \big\downarrow{\scriptstyle \pm 1} & & \big\downarrow{\scriptstyle \pm 1} \\ S^6 & \xrightarrow{\ell\omega} & S^3 \end{array}$$

It follows that $k\omega = \pm\ell\omega$, so that $k \equiv \pm\ell$ mod 12.

However, it is easy to prove that, for all primes p,

$$(E_1)_p \simeq (E_7)_p. \tag{1.4}$$

For we first observe that it is only necessary to localize at the primes $p = 2, 3$, since $\mathbb{Z}/12$ localizes to zero at other primes. Now if ω_p is the localization at p of ω, then ω_2 is of order 4 and ω_3 is of order 3. Thus

$$\omega_2 = -7\omega_2,$$
$$\omega_3 = 7\omega_3,$$

from which (1.4) may be deduced. Indeed it turns out that not only the total spaces E_1, E_7, but also the bundles themselves become (fibre-) homotopy equivalent on localizing at any prime.

The result (1.4) takes on a special interest when one observes that E_1 is the symplectic Lie group $Sp(2)$. E_7 is thus a manifold homotopically distinct from $Sp(2)$, but equivalent to $Sp(2)$ on localizing at any prime. We may also prove [6,8,9] that

$$Sp(2) \times S^3 = E_7 \times S^3, \text{ and} \tag{1.5}$$

$$Sp(2) \times Sp(2) \simeq E_7 \times E_7. \tag{1.6}$$

Either of these relations shows that E_7 is a Hopf manifold. Stasheff, using Zabrodsky's methods, showed [10] that E_7 has the homotopy type of a topological group G. However, we know that E_7 is not a Lie group. Nor indeed is G; thus G must be infinite-dimensional, since, were it finite-dimensional, it would have to be a manifold and therefore, according to the solution to Hilbert's Fifth Problem, it would admit the structure of a Lie group.

It is interesting to observe in this example that we obtain, by localization techniques, results (like (1.5)) which make no mention of localization. A further such result is, then, that the homotopy analog of Hilbert's Fifth Problem has a negative solution.

In this paper we will be concerned exclusively to construct the localization and to prove the most basic theorem giving the equivalent homotopy and homology characterizations. We will first do this in the homotopy category of 1-connected CW-complexes and will then proceed to generalize to the homotopy category of nilpotent CW-complexes. The generalization will be preceded by a section giving the basic definition and properties of nilpotent spaces. The reader only interested in the 1-connected case should find the section devoted to that case quite self-contained apart from the definition of localization of abelian groups. For the basic notions of localization of abelian and nilpotent groups, the reader is referred to [2,3].

2. Localization of 1-connected CW-complexes

We work in the pointed homotopy category H_1 of 1-connected CW-complexes. If $X \in H_1$, and if P is a family of primes, we say that X is P-local if the homotopy groups of X are all P-local abelian groups. We say that $f: X \to Y$ in H_1 P-localizes X if Y is P-local and*

$$f^*: [Y,Z] \cong [X,Z]$$

for all P-local $Z \in H_1$. Of course this universal property of f characterizes it up to canonical equivalence: if $f_i: X \to Y_i$, $i = 1,2$, both P-localize X then there exists a unique equivalence $h: Y_1 \simeq Y_2$ in H_1 with $hf_1 = f_2$. We will prove

Theorem 2A

Every X in H_1 admits a P-localization.

Theorem 2B

Let $f: X \to Y$ in H_1. Then the following statements are equivalent:

(i) f P-localizes X;

(ii) $\pi_n f: \pi_n X \to \pi_n Y$ P-localizes for all $n \geq 1$;

(iii) $H_n f: H_n X \to H_n Y$ P-localizes for all $n \geq 1$.

We will prove Theorems 2A, 2B simultaneously. We recall that a homomorphism $\phi: A \to B$ of abelian groups P-localizes if and only if B is P-local and ϕ is a P-isomorphism [2]; this latter condition means that the kernel and cokernel of ϕ belong to the Serre class C of abelian torsion groups with torsion prime to P. Thus to prove that (ii) $\iff$ (iii) in Theorem 2B above it suffices to prove the following two propositions.

Proposition 2.1

Let $Y \in H_1$. Then $\pi_n Y$ is P-local for all $n \geq 1$ if and if $H_n Y$ is P-local for all $n \geq 1$.

---

* We write, as usual, $[Y,Z]$ for $H_1(Y,Z)$, the set of pointed homotopy classes of maps from Y to Z.

Proposition 2.2

Let $f: X \to Y$ in $\mathcal{H}_1$. Then $\pi_n(f)$ is a P-isomorphism for all $n \geq 1$ if and only if $H_n(f)$ is a P-isomorphism for all $n \geq 1$.

Proof of 2.1 We first observe that H_nY is P-local for all $n \geq 1$ if and only if $H_n(Y; \mathbb{Z}/p) = 0$ for all $n \geq 1$ and all primes p disjoint from P. Now it was shown in [2] that if A is a P-local abelian group, so are the homology groups H_nA, $n \geq 1$. It now follows by induction on m, that if A is a P-local abelian group, so are the homology groups $H_n(A,m)$ of the Eilenberg-MacLane space $K(A,m)$. For we have a fibration $K(A,m-1) \to E \to K(A,m)$, with E contractible, from which we deduce that, if $H_n(A,m-1; \mathbb{Z}/p) = 0$ for all $n \geq 1$, then $H_n(A,m; \mathbb{Z}/p) = 0$ for all $n \geq 1$.

Now let $\ldots \to Y_m \to Y_{m-1} \to \ldots \to Y_2$ be the Postnikov decomposition of Y. Thus there is a fibration $K(\pi_mY,m) \to Y_m \to Y_{m-1}$, and $Y_2 = K(\pi_2Y,2)$. Thus, if we assume that π_nY is P-local for all $n \geq 1$, we may assume inductively that the homology groups of Y_{m-1} are P-local and we infer (again using homology with coefficients in $\mathbb{Z}/p$, with p disjoint from P) that the homology groups of Y_m are P-local. Since $Y \to Y_m$ is m-connected, it follows that H_nY is P-local for all $n \geq 1$.

To obtain the opposite implication, we construct the 'dual' Cartan-Whitehead decomposition

$$\ldots \to Y(m) \to Y(m-1) \to \ldots \to Y(2).$$

There is then a fibration $K(\pi_mY,m-1) \to Y(m+1) \to Y(m)$ and $Y(2) = Y$. Thus, if we assume that H_nY is P-local for all $n \geq 1$, we may assume inductively that the homology groups of $Y(m)$ are P-local. Since $\pi_mY(m) = \pi_mY$ and $Y(m)$ is (m-1)-connected, it follows that $\pi_mY \cong H_mY(m)$ and is P-local. Thus we infer (again using homology with coefficients in $\mathbb{Z}/p$, with p disjoint from P) that the homology groups of $Y(m+1)$ are P-local, so that the inductive step is complete and $\pi_nY \cong H_nY(n)$ is P-local.

Proof of 2.2 Since a P-isomorphism is an isomorphism mod C, where C is the class of abelian torsion groups with torsion prime to P, Proposition 2.2 is merely a special case of the classical Serre theorem.

We have thus proved that (ii) $\Leftrightarrow$ (iii) in Theorem 2B. We now prove that (ii) $\Leftrightarrow$ (i). The obstructions to the existence and uniqueness of a counterimage of $g: X \to Z$ under $f^*: [Y,Z] \to [X,Z]$ lie in $H^*(f;\pi_* Z)$. Now, given (ii) (or (iii)), $H_* f \in C$. Thus (i) follows from the universal coefficient theorem for cohomology and the following purely algebraic proposition [2].

Proposition 2.3

Let C be as in the proof of Proposition 2.2. Then, if $A \in C$ and B is P-local,

$$\text{Hom}(A,B) = 0, \quad \text{Ext}(A,B) = 0.$$

We now prove Theorem 2A. More specifically, we prove the existence of $f: X \to Y$ in $\mathcal{H}_1$ satisfying (iii). Since we know that (iii) $\Rightarrow$ (i), this will prove Theorem 2A. Our argument is facilitated by the following key observation.

Proposition 2.4

Let $\mathcal{U}$ be a full subcategory of $\mathcal{H}_1$, for whose objects X we have constructed $f: X \to Y$ satisfying (iii). Then the assignment $X \mapsto Y$ automatically yields a functor $L: \mathcal{U} \to \mathcal{H}_1$, for which f provides a natural transformation from the embedding $\mathcal{U} \subseteq \mathcal{H}_1$ to L.

Proof of 2.4 Let $g: X \to X'$ in $\mathcal{U}$. We thus have a diagram

$$\begin{array}{ccc} X & \xrightarrow{g} & X' \\ \downarrow f & & \downarrow f' \\ Y & & Y' \end{array} \tag{2.1}$$

in $\mathcal{H}_1$ with f, f' satisfying (iii). Since f satisfies (i) and Y' is P-local by Proposition 2.1, we obtain a unique (in $\mathcal{H}_1$) $h \in [Y,Y']$

making the diagram

(2.2)
$$\begin{array}{ccc} X & \xrightarrow{g} & X' \\ f\downarrow & & \downarrow f' \\ Y & \xrightarrow{h} & Y' \end{array}$$

commutative. It is now plain that the assignment $X \to Y$, $g \to h$ yields the desired functor L.

We exploit Proposition 2.4 to prove, by induction on n, that we may localize all n-dimensional CW-complexes in $\mathcal{H}_1$. If $n = 2$, then such a complex is merely a wedge of 2-spheres,

$$X = \bigvee_{\alpha} S^2$$

where α runs through some index set, and we define

$$Y = \bigvee_{\alpha} M(\mathbb{Z}_P, 2),$$

where $M(A,2)$ is the Moore space having $H_2M = A$. There is then an evident map $f: X \to Y$ satisfying (iii). Suppose now that we have constructed $f_0: X_0 \to Y_0$ satisfying (iii) if $\dim X_0 \leq n$, where $n \geq 2$, and let $\dim X = n + 1$, $X \in \mathcal{H}_1$. Then we have a cofibration

(2.3)
$$\vee S^n \xrightarrow{g} X^n \xrightarrow{i} X$$

By the inductive hypothesis and Proposition 2.4, we may embed (2.3) in the diagram

(2.4)
$$\begin{array}{ccccc} \vee S^n & \xrightarrow{g} & X^n & \xrightarrow{i} & X \\ \downarrow f_1 & & \downarrow f_0 & & \\ \vee M(\mathbb{Z}_P, n) & \xrightarrow{h} & Y_0 & \xrightarrow{j} & Y \end{array}$$

where f_0, f_1 satisfy (iii) and the square in (2.4) homotopy-commutes. Thus if j embeds Y_0 in the mapping cone Y of h, then we may complete (2.4) by $f: X \to Y$ to a homotopy-commutative diagram and it is then easy to prove (using the exactness of the localization of abelian groups) that $f: X \to Y$ also satisfies (iii). Thus we may construct $f: X \to Y$ satisfying (iii) if X is (n+1)-dimensional, and the inductive step is complete.

It remains to construct $f: X \to Y$ satisfying (iii) if X is infinite-dimensional. We have the inclusions

$$X^2 \subseteq X^3 \subseteq \ldots \subseteq X^n \subseteq X^{n+1} \subseteq \ldots$$

and may therefore construct

$$\begin{array}{ccc} X^n & \overset{i}{\subseteq} & X^{n+1} \\ \downarrow f^n & & \downarrow f^{n+1} \\ Y^{(n)} & \subseteq & Y^{(n+1)} \end{array} \tag{2.5}$$

where f^n, f^{n+1} satisfy (iii). We may even arrange that (2.5) is strictly commutative for each n. If we define $Y = \bigcup_n Y^{(n)}$, with the weak topology, then $Y \in \mathcal{H}_1$ and the maps f^n combine to yield a map $f: X \to Y$ which again obviously satisfies (iii). Thus we have proved Theorem 2A in the strong form that, to each X in $\mathcal{H}_1$, there exists $f: X \to Y$ in $\mathcal{H}_1$ satisfying (iii).

Finally, we complete the proof of Theorem 2B by showing that (i) $\Rightarrow$ (iii). Given $f: X \to Y$ which P-localizes X, let $f_0: X \to Y_0$ be constructed to satisfy (iii). Then $f_0: X \to Y_0$ also satisfies (i), from which one immediately deduces the existence of a homotopy equivalence $u: Y_0 \to Y$ with $uf_0 \simeq f$. It immediately follows that f also satisfies (iii).

Thus the proofs of Theorems 2A, 2B are complete.

3. Nilpotent spaces

It turns out that the category H_1 is not adequate for the full exploitation of localization techniques. This is due principally to the fact that it does not respect function spaces. We know, following Milnor, that if X is a (pointed) CW-complex and W a finite (pointed) CW-complex, then the function space X^W of pointed maps $W \to X$ has the homotopy type of a CW-complex. However its components will, of course, fail to be 1-connected even if X is 1-connected. However, it turns out that the components of X^W are nilpotent if X is nilpotent. Moreover, the category of nilpotent CW-complexes is suitable for homotopy theory (as first pointed out by E. Dror), and for localization techniques [11].

Definition 3.1 Let G be a group and let A be a G-module. Then we define the lower central G-series of A by

$$\Gamma^1 A = A; \qquad \Gamma^{n+1}A = \{a\text{-}xa,\ a \in \Gamma^n A,\ x \in G\},\ n \geq 1.$$

Moreover, A is said to be G-nilpotent, with nilpotency class c, where $c \geq 0$, if $\Gamma^c A \neq (0)$, $\Gamma^{c+1}A = (0)$. We also say that G operates nilpotently on A if A is G-nilpotent.

Definition 3.2 A connected CW-complex X is nilpotent if $\pi_1 X$ is nilpotent and operates nilpotently on $\pi_n X$ for every $n \geq 2$.

Let N be the homotopy category of nilpotent CW-complexes. Plainly $N \supseteq H_1$. Moreover, the simple CW-complexes are plainly in N; in particular N contains all connected Hopf spaces. We prove the following basic theorem.

Theorem 3.3

Let $F \xrightarrow{i} E \xrightarrow{f} B$ be a fibration of connected CW-complexes. Then $F \in N$ if $E \in N$.

Proof We exploit the classical result that the homotopy sequence of the fibration is a sequence of $\pi_1 E$-modules. We will prove that, if $\pi_n E$ is

$\pi_1 E$-nilpotent of class $\leq c$, then $\pi_n F$ is $\pi_1 F$-nilpotent of class $\leq c+1$. (A mild modification of the argument is needed to prove that if $\pi_1 E$ is nilpotent of class $\leq c$, then $\pi_1 F$ is nilpotent of class $\leq c+1$; we will deal explicitly with the case $n \geq 2$.)

We will need the fact that $\pi_1 E$ operates on $\pi_n B$ through f_*, and that the operation of $\pi_1 E$ on $\pi_n F$ is such that

(3.1) $$\xi \cdot \alpha = (i_* \xi) \cdot \alpha, \quad \xi \in \pi_1 F, \quad \alpha \in \pi_n F.$$

It will also be convenient to write I_F, I_E for the augmentation ideals of $\pi_1 F$, $\pi_1 E$. Then the statement that $\pi_n E$ is $\pi_1 E$-nilpotent of class $\leq c$ translates into

(3.2) $$I_E^c \cdot \pi_n E = (0).$$

Consider the exact sequence of $\pi_1 E$-modules

$$\cdots \longrightarrow \pi_{n+1} B \xrightarrow{\partial} \pi_n F \xrightarrow{i_*} \pi_n E \longrightarrow \cdots$$

and let $\xi \in I_F^c$, $\alpha \in \pi_n F$. Then $i_*(\xi \cdot \alpha) = (i_* \xi) \cdot i_*(\alpha) = 0$ by (3.2). Thus $\xi \cdot \alpha = \partial \beta$, $\beta \in \pi_{n+1} B$. Let $\eta \in \pi_1 F$. Then $\partial((i_* \eta - 1) \cdot \beta) = (i_* \eta - 1) \cdot \partial \beta$ $= (i_* \eta - 1) \xi \cdot \alpha = (\eta - 1) \xi \cdot \alpha$, by (3.1). But $(i_* \eta - 1) \cdot \beta = (f_* i_* \eta - 1) \cdot \beta = 0$, so $(\eta - 1) \xi \cdot \alpha = 0$. This shows that $I_F^{c+1} \cdot \pi_n F = (0)$, and thus the theorem is proved.

Now let W be a finite connected CW-complex and let X be a connected CW-complex. Let X^W be the function space of pointed maps $W \to X$ and let X_{fr}^W be the function space of free maps. Choose a map $g \in X^W$ $(g \in X_{fr}^W)$ as base point and let (X^W, g) $((X_{fr}^W, g))$ be the component of g.

<u>Theorem 3.4</u>

(i) (X^W, g) is nilpotent.

(ii) (X_{fr}^W, g) is nilpotent if X is nilpotent.

<u>Proof</u> We may suppose that W^0 is a point. Thus the assertions (i), (ii) are certainly true if W is 0-dimensional, and we will argue by induction

on the dimension of W. We will be content to prove (i). We have a cofibration

$$V \to W^n \to W^{n+1}$$

where V is a wedge of n-spheres, giving rise to a fibration

$$(X^{W^{n+1}}, g) \to (X^{W^n}, g_0) \to (X^V, o),$$

where $g: W^{n+1} \to X$ and $g_0 = g|W^n$. Our inductive hypothesis is that (X^{W^n}, g_0) is nilpotent, so that Theorem 3.3 establishes the inductive step.

Corollary 3.5

Let W be a finite CW-complex and $X \in N$. Then (X^W, g) and (X_{fr}^W, g) are nilpotent.

Proof Let $W_0, W_1, \ldots, W_d$ be the components of W, with $o \in W_0$. Then

$$X^W = X^{W_0} \times X_{fr}^{W_1} \times \ldots \times X_{fr}^{W_d}.$$

Since plainly a finite product of nilpotent spaces is nilpotent, it follows that (X^W, g) is nilpotent. Similarly (X_{fr}^W, g) is nilpotent.

Corollary 3.5 thus establishes (in view of Milnor's theorem) that we stay inside the category N when we take function spaces X^W with $X \in N$ and W finite.

We now proceed to give an important characterization of nilpotent spaces. Let X be a connected CW-complex and let

$$(3.3) \qquad \cdots \longrightarrow X_n \xrightarrow{p_n} X_{n-1} \longrightarrow \cdots \longrightarrow X_1 \longrightarrow o$$

be its Postnikov decomposition, so that p_n is a fibration with fibre $K(\pi_n X, n)$. We say that the Postnikov decomposition admits a principal refinement at stage n if p_n may be factored as a product of fibrations

$$(3.4) \qquad X_n = Y_c \xrightarrow{q_c} Y_{c-1} \longrightarrow \cdots \longrightarrow Y_1 \xrightarrow{q_1} Y_0 = X_{n-1},$$

where the fibre of q_i is an Eilenberg-MacLane space $K(G_i,n)$ and q_i is induced by a map $g_i\colon Y_{i-1} \to K(G_i,n+1)$, $1 \leq i \leq c$.

Theorem 3.6

> Let X be a connected CW-complex. Then the Postnikov decomposition of X admits a principal refinement at stage $n \geq 2$ (stage 1) if and only if $\pi_1 X$ operates nilpotently on $\pi_n X$ ($\pi_1 X$ is nilpotent).

Proof We will be content to give the argument for $n \geq 2$. Suppose first that we have the principal refinement (3.4). Then we may regard

$$Y_i \xrightarrow{q_i} Y_{i-1} \xrightarrow{g_i} K(G_i,n+1), \quad i = 1, \ldots, c,$$

as a fibration. Since $\pi_n Y_0 = (0)$, $\pi_1 X$ $(=\pi_1 Y_i, 1 \leq i \leq c)$ operates trivially on $\pi_n Y_0$. Thus, by repeated applications of the proof of Theorem 3.3, $\pi_1 X$ operates nilpotently on $\pi_n Y_c = \pi_n X_n = \pi_n X$.

Suppose conversely that $\pi_n X$ is $\pi_1 X$-nilpotent of class $\leq c$, and that we have factored $p_n\colon X_n \to X_{n-1}$ as

$$X_n \xrightarrow{r_i} Y_i \xrightarrow{s_i} X_{n-1},$$

where $r_i = q_{i+1} \cdots q_c$, each $q_j\colon Y_j \to Y_{j-1}$ being induced by $g_j\colon Y_{j-1} \to K(G_j,n+1)$, $G_j = \Gamma^j \pi_n X/\Gamma^{j+1}\pi_n X$, $i+1 \leq j \leq c$. If $i = c$, then $r_c = 1$, $s_c = p_n$. We suppose, moreover, that $q_{j*}\colon \pi_k Y_j \to \pi_k Y_{j-1}$ is the identity for $k < n$, that $q_{j*}\colon \pi_n Y_j \to \pi_n Y_{j-1}$ projects $\pi_n X/\Gamma^{j+1}\pi_n X$ onto $\pi_n X/\Gamma^j \pi_n X$ and that (by consequence) the homotopy groups of each Y_j vanish in dimensions $\geq n + 1$.

We attach $(n+1)$-cells to Y_i to kill the subgroup $\Gamma^i \pi_n X/\Gamma^{i+1}\pi_n X$ of $\pi_n Y_i = \pi_n X/\Gamma^{i+1}\pi_n X$; let Z' be the resulting space. We then attach $(n+2)$-cells, $(n+3)$-cells, ... , to kill $\pi_{n+1}, \pi_{n+2}, \ldots,$ so that we have finally embedded Y_i in a space Z, such that the effect on the homotopy groups occurs only in dimension n, where the induced homomorphism is the projection $\pi_n X/\Gamma^{i+1}\pi_n X \twoheadrightarrow \pi_n X/\Gamma^i \pi_n X$. Replace the inclusion $Y_i \subseteq Z$ by a fibration $q_i: Y_i \to Y_{i-1}$. Then the fibre of q_i is $K(G_i,n)$, where

$$G_i = \Gamma^i \pi_n X / \Gamma^{i+1} \pi_n X,$$

so that $\pi_1 X = \pi_1 Y_{i-1}$ operates trivially on G_i. It then follows, of course, that $\pi_1 X$ operates trivially on the homology of $K(G_i,n)$, so that q_i is induced by $g_i: Y_{i-1} \to K(G_i,n+1)$ -- we take the (negative) transgression of the fundamental class in the fibre. Moreover, it is clear that the map q_i is n-connected, so that an easy obstruction argument shows that s_i factors, up to homotopy, uniquely, as $s_i = s_{i-1}q_i$,

$$Y_i \xrightarrow{q_i} Y_{i-1} \xrightarrow{s_{i-1}} X_{n-1}.$$

Thus we continue until we have factored p_n as $s_0 q_1 \cdots q_c$, $s_0: Y_0 \to X_{n-1}$, with all the fibre maps q_i induced. However, it is now obvious that s_0 is a homotopy equivalence, so that we have proved that the Postnikov decomposition of X admits a principal refinement at stage n.

We would say that the Postnikov system of X admits a principal refinement if it admits a principal refinement at stage n for every $n \geq 1$. We then have the evident

Corollary 3.7

Let X be a connected CW-complex. Then X is nilpotent if and only if its Postnikov system admits a principal refinement.

We point out that the simple spaces are identified, by the correspondence implicit in this corollary, with those spaces whose Postnikov system is itself principal.

4. Localization of nilpotent complexes

In this section we extend Theorems 2A and 2B from the category H_1 to the category N. To do so we need, of course, to have the notion of the localization of nilpotent groups. This notion, together with the relevant properties, is to be found in [2,3], but we repeat the definition here for the reader's convenience. It will readily be seen that we are generalizing the localization of abelian groups in a very natural way.

Definition 4.1 A nilpotent group G is P-local if the function $x \mapsto x^p$, $x \in G$, is bijective for all primes p disjoint from P. A homomorphism $\phi: G \to K$ of nilpotent groups P-localizes if K is P-local and

$$\phi^*: \mathrm{Hom}(K,L) \to \mathrm{Hom}(G,L)$$

is bijective for all P-local nilpotent groups L.

Theorem 4.2

Every nilpotent group admits a P-localization.

Definition 4.3 Let $X \in N$. Then X is P-local if $\pi_n X$ is P-local for all $n \geq 1$. A map $f: X \to Y$ in N P-localizes if Y is P-local and

$$f^*: [Y,Z] \cong [X,Z]$$

for all P-local Z in N.

We now come to the main theorems of the paper.

Theorem 4A

Every X in N admits a P-localization.

Theorem 4B

Let $f: X \to Y$ in N. Then the following statements are equivalent:

(i) f P-localizes X;

(ii) $\pi_n f: \pi_n X \to \pi_n Y$ P-localizes for all $n \geq 1$;

(iii) $H_n f: H_n X \to H_n Y$ P-localizes for all $n \geq 1$.

The pattern of proof of these theorems will closely resemble that of Theorems 2A, 2B. However, an important difference is that the construction of a localization does not proceed cellularly, as in the 1-connected case, but via a principal refinement of the Postnikov system.

We first prove that (ii) ⇒ (iii) in Theorem 4B. We need a series of lemmas.

Lemma 4.4

If π acts nilpotently on A, then π acts nilpotently on $H_n(A,m)$, $n \geq 0$.

Proof Let $0 = \Gamma^{c+1}A \subseteq \Gamma^{c}A \subseteq \ldots \subseteq \Gamma^{1}A = A$ be the lower central π-series of A (see Section 3), and write $A_i = \Gamma^{i}A$ for convenience. Note that each A_i is a nilpotent π-module, of class less than that of A if $i \geq 2$. Moreover, π acts trivially on A_i/A_{i+1}. We have a spectral sequence of π-modules,

$$E_2^{pq} = H_p(A_i/A_{i+1},m;\ H_q(A_{i+1},m)),$$

converging (finitely) to the graded group associated with $H_n(A_i,m)$, suitably filtered. If we assume inductively that π operates nilpotently on $H_q(A_{i+1},m)$, it operates nilpotently on E_2^{pq}, whence it readily follows that π operates nilpotently on $H_n(A_i,m)$, completing the inductive step.

Lemma 4.5

Let $X \in N$ and let $\pi = \pi_1 X$. Then π operates nilpotently on $H_n(\tilde{X})$ where $\tilde{X}$ is the universal cover of X.

Proof Consider the Postnikov system of $\tilde{X}$. We have a fibration $K(\pi_m X,m) \to \tilde{X}_m \to \tilde{X}_{m-1}$, $m \geq 2$, where $\tilde{X}_1 = o$. Thus we may suppose inductively that π operates nilpotently on the homology of $K(\pi_m X,m)$. We have a spectral sequence of π-modules,

$$E_2^{pq} = H_p(\tilde{X}_{m-1};\ H_q(\pi_m X,m)),$$

converging (finitely) to the graded group associated with $H_n\tilde{X}_m$, suitably filtered. We see immediately that π operates nilpotently on

E_2^{pq}, whence it readily follows that π operates nilpotently on $H_n\tilde{X}_m$. This completes the inductive step. Since $\tilde{X} \to \tilde{X}_m$ is m-connected, the conclusion of the lemma follows.

Now, if G is a nilpotent group operating on A, and if the action is nilpotent, then, as shown in [3], there is a well-defined induced operation of G_P, the P-localization of G on A_P, the P-localization of A. Moreover, for these actions,

$$(\Gamma^i A)_P = \Gamma^i A_P.$$

Lemma 4.6

The induced map $H_n(G;A) \to H_n(G_P;A_P)$ is localization, $n \geq 0$.

Proof We argue by induction on the nilpotency class of A. If G operates trivially on A, then this follows from the universal coefficient theorem in homology and the basic result in [ ] that

(4.1) $$H_nG \to H_nG_P \text{ localizes, } n \geq 1.$$

We now consider the commutative diagram

$$\begin{array}{ccccc} \Gamma^{i+1}A & \rightarrowtail & \Gamma^i A & \twoheadrightarrow & \Gamma^i A/\Gamma^{i+1}A \\ \downarrow & & \downarrow & & \downarrow \\ \Gamma^{i+1}A_P & \rightarrowtail & \Gamma^i A_P & \twoheadrightarrow & \Gamma^i A_P/\Gamma^{i+1}A_P \end{array}$$

where the vertical arrows are localization. This induces the commutative diagram

$$\begin{array}{ccccccccc} \ldots H_{n+1}(G;\Gamma^i A/\Gamma^{i+1}A) & \to & H_n(G;\Gamma^{i+1}A) & \to & H_n(G;\Gamma^i A) & \to & H_n(G;\Gamma^i A/\Gamma^{i+1}A) & \to & H_{n-1}(\cdot)\to\ldots \\ \downarrow e_1 & & \downarrow e_2 & & \downarrow e_3 & & \downarrow e_4 & & \downarrow e_5 \\ \ldots H_{n+1}(G_P;\Gamma^i A_P/\Gamma^{i+1}A_P) & \to & H_n(G_P;\Gamma^{i+1}A_P) & \to & H_n(G_P;\Gamma^i A_P) & \to & H_n(G_P;\Gamma^i A_P/\Gamma^{i+1}A_P) & \to & H_{n-1}(\cdot)\to\ldots \end{array}$$

Here we know that e_1, e_4 localize and we may assume inductively that e_2, e_5 localize. It thus follows that e_3 localizes. This completes the inductive step and establishes the lemma.

We are now ready to prove that (ii) ⇒ (iii) in Theorem 4B. Let $\tilde{X}$, $\tilde{Y}$ be the universal covers of X, Y so that we have a diagram

(4.2)
$$\begin{array}{ccccc} \tilde{X} & \longrightarrow & X & \longrightarrow & K(\pi_1 X,1) \\ \downarrow{\scriptstyle \tilde{f}} & & \downarrow{\scriptstyle f} & & \downarrow{\scriptstyle f_1} \\ \tilde{Y} & \longrightarrow & Y & \longrightarrow & K(\pi_1 Y,1) \end{array}$$

Since $\tilde{f}$ induces localization in homotopy, it induces localization in homology by Theorem 2B. Moreover, we obtain from (4.2) a map of spectral sequences which is, at the E_2-level,

(4.3) $$H_p(\pi_1 X;H_q\tilde{X}) \to H_p(\pi_1 Y;H_q\tilde{Y}).$$

By Lemma 4.5 $\pi_1 X$ operates nilpotently on $H_q\tilde{X}$ and $\pi_1 Y$ operates nilpotently on $H_q\tilde{Y}$. We thus infer from Lemma 4.6, together with (4.1) if $q = 0$, that (4.3) is localization unless $p = q = 0$. Passing through the spectral sequences and the appropriate filtrations of $H_n\tilde{X}$, $H_n\tilde{Y}$, we infer that $H_n f$ localizes if $n \geq 1$.

Now let (i') be the statement $f^*\colon [Y,Z] \cong [X,Z]$ for all P-local Z in N. Note that this statement differs from (i) only in not requiring that Y be P-local. We prove that (iii) $\Rightarrow$ (i'). This will, of course, imply that (ii) $\Rightarrow$ (i).

If Z is P-local nilpotent, then we may find a principal refinement of its Postnikov system. Moreover this principal refinement will be such that the fibre at each stage is a space $K(A,n)$, where A is P-local abelian. For, as we saw in the proof of Theorem 3.6,

$$A = \Gamma^i \pi_n Z/\Gamma^{i+1}\pi_n Z$$

for some i, and it is easy to see, [3], that $\Gamma^i B$ is P-local if B is P-local. Given $g\colon X \to Z$, the obstructions to the existence and uniqueness of a counterimage to g under f^* will thus lie in the groups $H^*(f;A)$ and, as in the corresponding argument in the 1-connected case (note that we have trivial coefficients here, too), these groups will vanish if f induces P-localization in homology.

Next we proceed to prove Theorem 4A, via a key observation playing the role of Proposition 2.4.

Proposition 4.7

Let $\mathcal{U}$ be a full subcategory of $\mathcal{N}$, for whose objects X we have constructed $f: X \to Y$ satisfying (ii). Then the assignment $X \mapsto Y$ automatically yields a functor $L: \mathcal{U} \to \mathcal{N}$, for which f provides a natural transformation from the embedding $\mathcal{U} \subseteq \mathcal{N}$ to L.

Proof of 4.7 Let $g: X \to X'$ in $\mathcal{U}$. We thus have a diagram

$$\begin{array}{ccc} X & \xrightarrow{g} & X' \\ \downarrow f & & \downarrow f' \\ Y & & Y' \end{array} \qquad (4.4)$$

in $\mathcal{N}$, where f, f' satisfy (ii). Then f satisfies (i) and Y' is P-local, so that there exists a unique h in $\mathcal{N}$ such that the diagram

$$\begin{array}{ccc} X & \xrightarrow{g} & X' \\ \downarrow f & & \downarrow f' \\ Y & \xrightarrow{h} & Y' \end{array} \qquad (4.5)$$

commutes. It is now plain that the assignment $X \mapsto Y$, $g \mapsto h$ yields the desired functor L.

We now exploit Proposition 4.7 to prove Theorem 4A. We first consider spaces X in $\mathcal{N}$ yielding a finite refined principal Postnikov system and, for those, we argue by induction on the height of the system. Thus we may assume that we have a principal (induced) fibration

$$K(G,n) \to X \to X', \qquad (4.6)$$

where G is abelian even if $n = 1$, and we may suppose that we have constructed $f': X' \to Y'$ satisfying (ii). (The induction starts with $X' = o$.) Since (4.6) is induced, we may, in fact, assume a fibration

$$X \longrightarrow X' \xrightarrow{g} K(G,n+1)$$

Now we may certainly localize $K(G,n+1)$; we obtain $K(G_P,n+1)$, where

G_P is the localization of G and so, by Proposition 4.7, we have a diagram

$$\begin{array}{ccccc} X & \longrightarrow & X' & \xrightarrow{g} & K(G,n+1) \\ & & \downarrow{\scriptstyle f'} & & \downarrow{\scriptstyle e} \\ & & Y' & \xrightarrow{h} & K(G_P,n+1) \end{array}$$

Let Y be the fibre of h. There is then a map $f\colon X \to Y$ rendering the diagram

$$\begin{array}{ccccc} X & \longrightarrow & X' & \xrightarrow{g} & K(G,n+1) \\ \downarrow{\scriptstyle f} & & \downarrow{\scriptstyle f'} & & \downarrow{\scriptstyle e} \\ Y & \longrightarrow & Y' & \xrightarrow{h} & K(G_P,n+1) \end{array}$$

commutative in N and a straightforward application of the exact homotopy sequence shows that f satisfies (ii).

It remains to consider the case in which the refined principal Postnikov system of X has infinite height (this is, of course, the 'general' case!). Thus we have principal fibrations

(4.8) $$\cdots \longrightarrow X_i \xrightarrow{g_i} X_{i-1} \longrightarrow \cdots \longrightarrow o$$

and there is a weak homotopy equivalence $X \to \varprojlim X_i$.

Now we may apply the reasoning already given to embed (4.8) in the diagram, commutative in N,

(4.9) $$\begin{array}{ccccccccc} \cdots & \longrightarrow & X_i & \xrightarrow{g_i} & X_{i-1} & \longrightarrow & \cdots & \longrightarrow & o \\ & & \downarrow{\scriptstyle f_i} & & \downarrow{\scriptstyle f_{i-1}} & & & & \\ \cdots & \longrightarrow & Y_i & \xrightarrow{h_i} & Y_{i-1} & \longrightarrow & \cdots & \longrightarrow & o \end{array}$$

where each f_i satisfies (ii). Moreover, we may suppose that each h_i is a fibre map. Let Y be the geometric realization of the singular complex of $\varprojlim Y_i$. Then there is a map $f\colon X \to Y$ such that the diagram

$$\begin{array}{ccc} X & \longrightarrow & \varprojlim X_i \\ \downarrow{\scriptstyle f} & & \downarrow{\scriptstyle \varprojlim f_i} \\ Y & \longrightarrow & \varprojlim Y_i \end{array}$$

is homotopy-commutative. Moreover, the construction of (4.9) shows that the Y_i-sequence is again a refined principal Postnikov system, from which it readily follows that $\varprojlim f_i$ satisfies (ii). So therefore does f, and f is in $\mathcal{N}$. Thus we have completed the proof of Theorem 4A in the stronger form that there exists, for each X in $\mathcal{N}$, a map $f: X \to Y$ in $\mathcal{N}$ satisfying (ii).

The proof that (i) $\Rightarrow$ (ii) proceeds exactly as in the easier case of the category $\mathcal{H}_1$. Thus we have established the following set of implications:

(4.10) (ii) $\Rightarrow$ (iii), (iii) $\Rightarrow$ (i'), (ii) $\Rightarrow$ (i), (i) $\Rightarrow$ (ii).

All that remains is to prove the following proposition, for then we will be able to infer, in fact, (iii) $\Rightarrow$ (i).

Proposition 4.8

If $Y \in \mathcal{N}$ and H_nY is P-local for every $n \geq 1$, then π_nY is P-local for every $n \geq 1$.

Proof To prove this, we invoke Dror's theorem, which we, in fact, reprove since it follows immediately from (4.10). Thus we consider the special case $P = \Pi$, where Π is the collection of all primes. Then a homomorphism of (nilpotent, abelian) groups Π-localizes if and only if it is an isomorphism. Moreover, every space in $\mathcal{N}$ is Π-local, so that, in this special case, the distinction between (i') and (i) disappears. Thus (4.10) implies, in particular, the equivalence of (ii) and (iii) for $P = \Pi$, which is Dror's theorem.

Now we prove Proposition 4.8. We construct $f: Y \to Z$ satisfying (ii). It thus also satisfies (iii); but H_nY is P-local, so that f induces an isomorphism in homology. By Dror's theorem, f induces an isomorphism in homotopy. However, the homotopy of Z is P-local, so that Proposition 4.8 is proved, and, with it, the proof of Theorems 4A and 4B is complete.

Remarks

1. Of course, we do not need the elaborate machinery assembled in this section to prove Dror's theorem. In particular, Theorem 4A is banal for $P = \Pi$, since, then, the identity $X \to X$ Π-localizes!

2. The fact that we have both the homotopy criterion (ii) and the homology criterion (iii) for the localizing map f enables us to derive some immediate conclusions. For example, we may use (ii) to prove ([4]):

Theorem 4.9

If X is nilpotent and W finite, and if $f\colon X \to Y$ localizes, then $f^W\colon X^W \to Y^W$ localizes.

Similarly, we use (ii) to prove

Theorem 4.10

If $F \to E \to B$ is a fibre sequence in N, then so is

$$F_P \to E_P \to B_P,$$

where X_P is the P-localization of X.

Finally, we use (iii) to prove

Theorem 4.11

If $U \to V \to W$ is a cofibre sequence in N, then so is

$$U_P \to V_P \to W_P,$$

where X_P is the P-localization of X.

3. An important reason for the difference between the proofs of Theorem 2B and Theorem 4B is that, in H_1, we can construct the localization cellularly, whereas in N we construct it homotopically. It would be very interesting to know whether the localization can be constructed cellularly in N.

References

[1] A. K. Bousfield and D. M. Kan: Homotopy limits, completions and localizations. Lecture Notes in Mathematics 304, Springer (1972).

[2] P. J. Hilton: Localization and cohomology of nilpotent groups. Math. Zeit. (1973) (to appear).

[3] P. J. Hilton: Remarks on the localization of nilpotent groups. Comm. Pure and Applied Math. (1973) (to appear).

[4] P. J. Hilton, G. Mislin and J. Roitberg: Homotopical localization. Proc. Lond. Math. Soc. 3, XXVI (1973), 693-706.

[5] P. J. Hilton, G. Mislin and J. Roitberg: H-spaces of rank 2 and non-cancellation phenomena. Inv. Math. 16 (1972), 325-334.

[6] P. J. Hilton and J. Roitberg: On principal S^3-bundles over spheres. Ann. of Math. 90 (1969), 91-107.

[7] M. Mimura, G. Nishida and H. Toda: Localization of CW-complexes and its applications. J. Math. Soc. Japan 23 (1971), 593-624.

[8] G. Mislin: The genus of an H-space. Lecture Notes in Mathematics 249, Springer (1971), 75-83.

[9] A. Sieradski: Square roots up to homotopy type. Amer. J. Math. 94 (1972), 73-81.

[10] J. Stasheff: Manifolds of the homotopy type of (non-Lie) groups. Bull. Amer. Math. Soc. 75 (1969), 998-1000.

[11] D. Sullivan: Geometric topology, part I: Localization, periodicity and Galois symmetry. MIT (June 1970) (mimeographed notes).

[12] A. Zabrodsky: Homotopy associativity and finite CW-complexes. Topology 9 (1970), 121-128.

TOPOLOGICAL METHODS IN ECONOMICS: GENERAL EQUILIBRIUM THEORY

Erwin Klein

Dalhousie University, Halifax

1. General Equilibrium Theory

Traditionally, General Equilibrium Theory (GET) has been the branch of economics concerned with the explanation of prices (Walras, 1874-77) and the properties of optimal economic states (Pareto, 1906). It remained until the beginning of the fifties an essentially calculus-based discipline.

It is during the 1950's that GET experiences its great formal transformation by adopting its current topology-based approach.

The scope of GET has since enlarged considerably. Besides the classical equilibrium-existence problem whose solution has usually been attempted by means of algebraic-topology methods (fixed point theorems), new areas of research have been opened: The problem of the finiteness of the set of equilibria, with mathematical tools borrowed from differential topology; the question of the properties of solutions and concepts, alternative to the classical of equilibrium, and of the relations between them, with mathematical tools which include topology, measure theory and the theory of games, stability theory, including tools standard in dynamic mathematical analysis. Finally, besides the very developed theory of a perfectly competitive economy, the last few years have seen the serious beginning of a theory of an imperfectly competitive economy.

This paper only aims at describing the alternative topological "set-ups" in which certain economic theoretical problems are tackled. To provide some unity to our discussion, the three cases which will be analyzed belong to the "equilibrium-existence" problem.

The three models to be described constitute severe simplifications of reality: (a) perfect competition is assumed (no individual agent may influence prices; it only takes these as given action parameters), (b) households and firms are the only decision-making agents of the economy (government, public bodies, and thus also collective goods are not included in the discussion), and (c) commodities are directly exchanged against commodities (no special place is recognized to money). They will, however, provide some deep insights into the working of an economy. And, above all, they will serve to illustrate alternative topological "set ups" as currently used in the field. To show this, after all, is the main objective of this paper.

2. Finite Economies

A most natural picture is that of an economy with a finite number of commodities (indexed by $i = 1,\dots,n$), a finite number of firms (indexed by $f = 1,\dots,F$), and a finite number of households (indexed by $h = 1,\dots,H$).

A commodity is any good or service well identified by three characteristics: (a) physical properties, (b) location of delivery, (c) time of delivery. A commodity bundle is an n-tuple of real numbers in the commodity space R^n, with the ith component representing the quantity in which the ith commodity is included

The hth household or consumer is represented by a consumption possibility set X_h R^n, totally preordered by a preference-indifference relation $\precsim_h$, together with a vector of initial resources, $\bar{x}_h \in R^n$. A point $x_h \in X_h$ is called a consumption programme, where positive components are inputs (consumption proper) and negative components are outputs (supply of labour services) The aggregate consumption possibility set of the economy is

$$X = \sum_{h=1}^{H} X_h \ ;$$

a total consumption programme is $x = \sum_{h=1}^{H} x_h$, where, clearly, $x \in X$; the total initial resources are given by

$$\bar{x} = \sum_{h=1}^{H} \bar{x}_h .$$

The fth firm or producer is represented by a production possibility set $Y_f \subset R^n$. A production programme is a point $y_f \in Y_f$, where negative components represent inputs (raw materials, labour, etc.) and positive components outputs (production proper). The total production possibility set of the economy is

$$Y = \sum_{f=1}^{F} Y_f ,$$

a total production programme

$$y = \sum_{f=1}^{F} y_f \in Y$$

With each commodity there is associated a nonnegative real number, its price. If Ω stands for the nonnegative orthant of R^n, then we shall represent a normalized price system by an n-tuple

$$p \in P = \{p \in \Omega \mid \sum_{i=1}^{n} p_i = 1\}.$$

The profit of f from producing y_f at p is the inner product $\pi_f = py_f$; total profit of the economy is thus

$$\Pi = py = \sum_{f=1}^{F} py_f .$$

The inner products px_h, px, $p\bar{x}_h$, $p\bar{x}$ stand for the consumption expenditure value of h of x_h at p, the value of total consumption expenditure, the value of the initial endowment of h, the value of the total initial resources of the economy, respectively.

A possible submodel is that of a private ownership economy, which emerges if we assume that consumers own resources and firms, and that the hth consumer has a claim on the profit of the fth firm given by $\alpha_{hf} \geqq 0$, with

$$\sum_{h=1}^{H} \alpha_{hf} = 1 , \quad \forall f .$$

Under the above conditions, wealth of the hth household is

$$w_h = p\bar{x}_h + \sum_{f=1}^{F} \alpha_{hf}\, py_f .$$

A point $w \in R^H$ is called a wealth distribution. If p and w are given, then $(p,w) \in R^{H+n}$ is called a price-wealth pair A point $x - y - \bar{x} = z$ is an excess-demand The set of excess-demands is thus $X - Y - \{\bar{x}\} = Z$. We now have that h owns the vector $\bar{x}_h$.

With all elements as described so far, a private ownership finite economy is a system

$$E = (X_h , \preccurlyeq_h , Y_f , \bar{x} , \alpha_{hf})$$

with commodity space R^n, resources $\bar{x} = \sum_{h=1}^{H} \bar{x}_h$, and $h = 1,\ldots,H$, $f = 1,\ldots,F$. A comprehensive picture of the relevant literature is found in Arrow-Hahn (1971). Our description is mainly based on Debreu (1959).

(a) Set Properties and Mappings

The following assumptions for production sets are standard: A1: $\forall f$, Y_f is closed, convex, and contains 0; A2: $Y \cap \Omega = \{0\}$ (impossibility of free production); A3: $Y \cap (-Y) = \{0\}$ (irreversibility); A4: $-\Omega \subset Y$ (possibility of free disposal).

One can prove: if A1 and A3 hold, then Y is closed.

Define $B_f = \{p \in P \mid pY_f$ has a maximum$\}$. Then (profit-maximization principle) a supply correspondence for f is a mapping $s_f\colon B_f \to Y_f$ defined by $s_f(p) = \{y_f \in Y_f \mid py_f = \max pY_f\}$. An aggregate supply correspondence for the economy is defined by

$$s(p) = \sum_{f=1}^{F} s_f(p).$$

A profit function for f is a mapping $\pi_f\colon P \to R$ defined by $\pi_f(p) = \max pY_f$. A total profit function is now Π where

$$\Pi(p) = \sum_{f=1}^{F} \pi_f(p).$$

Standard assumptions for consumption sets are: A5: X_h is closed and convex $\forall h$; A6: X_h is lower bounded for $\forall h$; A7: $\forall h$, $\preccurlyeq_h$ is a total preordering such that: (a) $\forall x_h^o \in X_h$, $\{x_h \in X_h \mid x_h^o \preccurlyeq_h x_h\}$ and $\{x_h \in X_h \mid x_h \preccurlyeq_h x_h^o\}$ are closed in X_h (continuity),

(b) $\forall x_h^o \in X_h$, $\{x_h \in X_h \mid x_h^o \preccurlyeq_h x_h\}$ is convex (convexity), and

(c) X_h has no maximal element for $\preccurlyeq_h$ (nonsatiation).

One can prove: If A5 and A6 hold, then X is closed.

One can also prove: If A5 and A7 hold, then a continuous real-valued utility function (order-preserving mapping) $u_h: X_h \to R$ exists. If so, then a consumer can also be represented by u_h together with $\bar{x}_h$.

Define $E_h = \{(p,w) \in R^{H+n} \mid x_h \in X_h$ such that $px_h \leqq w_h\}$. Then the mapping $\gamma_h: E_h \to X_h$ defined by $\gamma_h(p,w) = \{x_h \in X_h \mid px_h \leqq w_h\}$ is the budget correspondence of h. If, additionally, we define

$$E'_h = \{(p,w) \in E_h \mid \gamma_h(p,w) \text{ has maximal element for } \precsim_h\},$$

a demand correspondence for h (preference-satisfaction principle) is a mapping $d_h: E'_h \to X_h$ defined by $d_h(p,w) = \{x_h \in \gamma_h(p,w) \mid x_h$ is maximal element in $\gamma_h(p,w)$ for $\precsim_h\}$ or, alternatively, $d_h(p,w) = \{x_h \in \gamma_h(p,w) \mid x_h$ is maximizer of u_h on $\gamma_h(p,w)\}$. An aggregate demand correspondence can now be defined by

$$d(p,w) = \sum_{h=1}^{H} d_h(p,w).$$

Observe that once $\bar{x}_h$ is introduced into the scene, d_h becomes a function of p alone for $w_h = p\bar{x}_h + \sum_f \alpha_{hf}\, py_f$. The same holds, naturally, for d.

A state of the economy is a point $((x_h),(y_f))$ in $R^{n(H+F)}$. It is said to be a market equilibrium if it belongs to the linear manifold $M = \{((x_h),(y_f)) \mid z = x - y - \bar{x} = 0\}$. It is attainable if it belongs to $A = \{((x_h),(y_f)) \mid x_h \in X_h\ \forall h,\ y_f \in Y_f\ \forall f,\ z = 0\}$. The set A is clearly

$$M \cap \left(\left(\prod_{h=1}^{H} X_h\right) \times \left(\prod_{f=1}^{F} Y_f\right)\right).$$

(b) Some Important Theorems

If X_h is lower bounded for $\leqq$, $\forall h$, so is X.

One can prove: If X is lower bounded for $\leqq$, and A1, A2, and A3 hold, then A is bounded.

Since A is bounded, so is the consumption attainable set $\hat{X}_h \subset X_h$, $\forall h$, and the production attainable set $\hat{Y}_f \subset Y_f$, $\forall f$. In the analysis, we may either replace $X_h, \forall h$, and $Y_f, \forall f$ by the compact sets $\hat{X}_h$ and $\hat{Y}_f$, or simply assume that X_h and Y_f are compact $\forall h, \forall f$.

If X_h is compact and convex $\forall h$, and if we assume A8: $\forall h$, $w_h > \inf pX_h$, then $\gamma_h(p,w)$ is continuous at (p,w), $\forall h$.

Define $f_f\colon Y_f \to R$ such that $f_f(y_f) = py_f$, which is continuous for given $p \in P$, and $F_f\colon P \to Y$ such that $F_f(p) = Y_f$ which is trivially continuous. We can then prove: $s_f\colon P \to Y_f$ with

$$s_f(p) = \{y_f \in Y_f \mid py_f = \max pY_f\}$$

is upper semi-continuous, and $\pi_f\colon P \to R$ with $\pi_f(p) = \max pY_f$ is continuous. With similar arguments, using the indirect utility function $\nu_h\colon E_h' \to R$ with $\nu_h(p,w) = \max u_h(\gamma_h(p,w))$ one can prove the upper semi-continuity of d_h and the continuity of ν_h. The upper semi-continuity of s and d follows. The same then holds for $e\colon P \to Z$, the excess demand correspondence.

The convex-valuedness of s, d, and e is, in general, easy to show.

(c) The Equilibrium Existence Problem

An equilibrium for E is a triplet $((x_h^*),(y_f^*),p^*)$ such that

(i) $\forall h$, x_h^* is maximal element of $\{x_h \in X_h \mid p^*x_h \leqq p^*\bar{x}_h + \Sigma_f\, \alpha_{hf}p^*y_f^*\}$ for $\precsim_h$;

(ii) $\forall f$, y_f^* is maximizer of p^*Y_f on Y_f ;

(iii) $x^* - y^* = \bar{x}$.

Since an equilibrium consumption has to satisfy the wealth constraint for every household h, we have

$$\sum_h px_h \leqq \sum_h p\bar{x}_h + \sum_h \sum_f \alpha_{hf}\, py_f, \quad \text{i.e.} \quad px \leqq p\bar{x} + py, \quad \text{or} \quad p\,z \leqq 0.$$

In general, $pz(p) \leqq 0$ (Walras Law). Since $p \in P$, we necessarily must have $z(p) \leqq 0$.

The essential piece in the proof of the existence of an equilibrium is thus the proof that $\exists\, p \in P$ such that $z(p) \leq 0$ (this includes, then, the case $z(p) = 0$) or, what is the same, that $\exists\, p \in P$ such that $z(p) \cap (-\Omega) \neq \emptyset$. This is done by means of the

<u>Excess-Demand Theorem</u>. Let $Z \subset R^n$ be compact. If the mapping $z\colon P \to Z$ is an upper semi-continuous correspondence such that $\forall p \in P$ the image set $z(p)$ is non-empty, convex, and satisfies Walras Law ($pz(p) \leq 0$), then $\exists\, p \in P$ such that $z(p) \cap (-\Omega) \neq \emptyset$.

The strategy of the proof consists in constructing a mapping $\Psi\colon P \times Z \to P \times Z$ defined by $\eta(z) \times z(p)$, where $\eta(z) = \{p \in P \mid p$ maximizing pz on $P\}$. It can be proved that Ψ has a fixed point (Kakutani), which satisfies the equilibrium definition. Additionally, if $p >> 0$, then $z(p) = 0$.

The second step in the existence proof is done by showing that the described economy satisfies the conditions for the application of the crucial Excess-Demand Theorem. For instance:

<u>Equilibrium Existence Theorem</u>. If E satisfies <u>A1</u> - <u>A8</u>, then it has an equilibrium.

For an E with the mentioned properties it can also be proved that (a) the allocation $((x_h^*),(y_f^*))$ which corresponds to the equilibrium triplet $((x_h^*),(y_f^*),p^*)$ is a maximal element (Pareto optimum) of the partially preordered set A; (b) for an economy $E = (X_h, \precsim_h, Y_f, \bar{x})$ -- note that we do not call it "private ownership", since the ownership of the firms and resources is not specified -- $\exists\, p^*$ for which the maximal element $((x_h^*),(y_f^*))$ is an equilibrium. If we want to extend the theorem from E to E, then to make $((x_h^*),(y_f^*))$ an equilibrium allocation may not only require a suitable $p^* \in P$ but also a redistribution of the initial resources $\bar{x}$.

3. Economies with a measure space of consumers

The motivation to study economies with a continuum of agents has four main aspects: (a) an atomless measure space appears to be the natural environment of economic agents who are supposed to strictly satisfy the perfectly competitive assumption about their negligible individual influence upon collective outcomes; (b) theorems relating alternative solutions for economies may be better formulated for continuum economies than by studying the problem for an infinite sequence of economies $(\mathcal{E}^k \mid k \in N)$, where N is the set of natural numbers and where k, the number of agents, tends to infinity; (c) if, under suitable convergence conditions, the economy with a measure space of agents can be shown to be the limit of a sequence of finite economies, then perhaps properties of the converging finite economies can be derived from the properties of the "limit economy"; (d) finally, it is expected that insight into the properties of imperfectly competitive economies may be made accessible by this approach, by widening the analysis so as to include measure-space economies with atomless as well as atomic parts.

Pure exchange economies of this type have been studied by many authors, among others by Aumann (1964, 1966, 1973), Hildenbrand (1968), Kannai (1968), Schmeidler (1968), Vind (1964). In this paper we focus our attention on a model which includes a finite production sector besides the continuum of consumers, and which is due to Hildenbrand (1970).

(a) Basic Concepts and Definitions

The novel element in this model, as compared with the economy of the preceding section, is the definition of a measure space of consumers, a system $((H,\mathcal{H},\mu),X,\preccurlyeq,\bar{x})$.

A household or consumer is a point $h \in H$. $\mathcal{H}$ is a σ-algebra of subsets of H, where $C \in \mathcal{H}$ is called a coalition of consumers. Since μ is a σ-additive positive measure on $\mathcal{H}$, with $\mu(H) = 1$, $(H,\mathcal{H},\mu)$ is a measure space. To denote the fraction of the total of consumers in the coalition C we thus write $\mu(C)$.

Write $\mathcal{B}$ for the σ-algebra of Borel subsets of the finite commodity space R^n, $\mathcal{H}_\mu$ for the completion of $\mathcal{H}$ relative to μ, and

$\mathcal{H}_\mu \otimes \mathcal{B}$ for the product σ-algebra. Define the μ-measurable correspondence $X: H \to R^n$ with graph included in $\mathcal{H}_\mu \otimes \mathcal{B}$, and given by $G_X = \{(h,x) \in H \otimes R^n \mid x \in X(h)\}$. The set $X(h)$ is then called the consumption possibility set of $h \in H$.

It is possible to define a measurable preference function $\preccurlyeq$ if a total preordering $\preccurlyeq_h$ is defined $\forall h \in H$ such that

$$\{(h, x, x') \in H \otimes R^n \otimes R^n \mid x \preccurlyeq_h x'\}$$

is included in $\mathcal{H}_\mu \otimes \mathcal{B} \otimes \mathcal{B}$.

Let $\Psi: H \to R^n$ be a correspondence and call L_Ψ the set $\{f \mid f: H \to R^n$ is μ-integrable with $f(h) \in \Psi(h)$ a.e.$\}$. Then $f \in L_X$ denotes a consumption programme. Finally, as far as households are concerned, the initial distribution of resources is given by means of the μ-measurable function $\bar{x}: H \to R^n$.

Before describing profit distribution to households it is necessary to identify the (finite) production sector. Simply, a firm is a production possibility set $Y_f \subset R^n$, and we have $f = 1,\ldots,F$.

$\forall f$, define the share-distribution function A_f, where $A_f: \mathcal{H} \to [0,1]$ is a σ-additive measure such that $\mu(C) = 0$ implies $A_f(C) = 0$. By Radon-Nykodym, $\exists\, \alpha_f: H \to R$, nonnegative functions (densities of shares distribution) with $\int_H \alpha_f \, d\mu = 1$.

A "private ownership economy with a measure space of consumers" is now a system $\mathcal{E} = ((H,\mathcal{H},\mu),X,\preccurlyeq,\bar{x},Y_f,\alpha_f)$ where $f = 1,\ldots,F$, and the commodity space is R^n.

Wealth distribution is described by means of the μ-measurable function $w: H \to R$ defined by $w(h,p) = p\bar{x}(h) + \sum_f \alpha_f(h) \sup pY_f$, and where p is again a price system.

Accordingly, a budget correspondence can be given by

$$\gamma(h,p) = \{x \in X(h) \mid px \leq w(h,p)\},$$

and a demand correspondence by

$$d(h,p) = \begin{cases} \{x \in \gamma(h,p) \mid x^o \precsim_h x, \ \forall x^o \in \gamma(h,p)\} , \\ \qquad\qquad \text{if } \inf pX(h) < w(h,p) , \text{ and} \\ \gamma(h,p) , \quad \text{if } \inf pX(h) \geqq w(h,p) . \end{cases}$$

The aggregate demand correspondence easily follows:

$$d(p) = \int_H d(h,p)\, d\mu .$$

Nothing special, at this point, has to be said about supply correspondences.

For E, a quasi-equilibrium is a triplet (f^*, y^*, p^*) with $f^* \in L_X$, $y^* \in Y$, and $p^* \in R^n$ with $p^* > 0$ such that:

(i) a.e. in H :
$f^*(h)$ is maximal element in $\gamma(h,p)$ for $\precsim_h$;

(ii) $\forall f$:
$p^* y^*_f = \max p^* Y_f$;

(iii) $\int_H f^*\, d\mu = \sum_{f=1}^{F} y^*_f + \int_H \bar{x}\, d\mu$.

A quasi-equilibrium becomes an equilibrium (compare with Chapter 2. on Finite Economies) if $\inf pX(h) < w(h,p)$. A "collective assumption" which satisfies this -- an analog exists for finite economies, too -- is

$$(\bar{x}(h) + \text{int } AY) \cap X(h) \neq \emptyset , \quad \text{a.e. in } H ,$$

where int AY is the interior of the asymptotic cone of Y.

Standard assumptions on consumption possibility sets are:
A1: a.e. in H : $X(h)$ is closed and convex; A2: $\precsim_h$ is a total preordering, continuous (see Section 2.), and locally non-satiated (in every neighbourhood $\eta(x^o)$, $\exists x \in X(h) \cap \eta(x^o)$ with $x^o \precsim_h x$). If $h \in H$ belongs to an atom (atoms are thus not excluded) it is additionally required that $\precsim_h$ be convex (see Section 2.). It is also assumed that A3: $g : H \to R^n$, μ-integrable, such that $\forall h \in H$, $g(h) \leqq X(h)$, and that A4: $\bar{x}$ is μ-integrable.

For the production possibility sets we have A5: $0 \in Y_f$, $\forall f$;

A6: Y is closed and convex; A7: $Y \cap \Omega = \{0\}$ (impossibility of free production); A8: $Y \cap (-Y) = \{0\}$ (irreversibility); and A9: $-\Omega \subset Y$ (free disposal). An additional "collective" property required is
A10: $(\bar{x}(h) + AY) \cap X(h) \neq \emptyset$ (weak adequacy) or $(\bar{x}(h) + \text{int } AY) \cap X(h) \neq \emptyset$ (strong adequacy), a.e. in H.

Equilibrium (or quasi-equilibrium) existence proofs are attempted in a roundabout way, for which the concept of a "coalition production economy" is essential. We describe it next, reminding that $C \in \mathcal{H}$ is called a coalition.

(b) A coalition production economy

In a coalition economy, coalitions of consumers own the firms. It is necessary to specify, hence, the way in which every coalition $C \in \mathcal{H}$ owns the firms as well as the profit-distribution mechanism.

A coalition production economy is a system $\overset{o}{\mathcal{E}} = ((H,\mathcal{H},\mu),X,\preccurlyeq,\bar{x},\overset{o}{Y})$ where every symbol has the already mentioned meaning, and where $\overset{o}{Y}: \mathcal{H} \to R^n$, called the coalition set correspondence, is countably additive.

To every $C \in \mathcal{H}$ we assign a production possibility set Y_C such that $0 \in Y_C$. $\overset{o}{Y}$ is assumed to admit a Radon-Nykodym derivate (for suitable conditions see Debreu and Schmeidler (1970)), that is $\exists\, Y: H \to R^n$, a set-valued mapping, such that $Y_C = \int_C Y \, d\mu$. Also, every Y is assumed closed and convex-valued.

If Y is a Radon-Nykodym derivative of $\overset{o}{Y}$, then $\forall p$, $\exists\, \pi(p,h) = \sup pY(h)$, a.e. in H. It can be proved that

$$\forall C \in \mathcal{H}, \quad \int_C \sup Y \, d\mu = \int_C \pi(p,\cdot) \, d\mu = \Pi(p,C) .$$

$\Pi(p,C)$ is the coalition profit distribution. $\Pi(p,\cdot)$ is a countably additive measure such that $\mu(C) = 0$ implies $\Pi(p,C) = 0$. Then, $\pi(p,\cdot)$ is a Radon-Nykodym derivative of $\Pi(p,\cdot)$, called a profit function.

Wealth distribution is specified by the μ-measurable function $w(h,p) = p\bar{x}(h) + \sup pY(h)$.

A quasi-equilibrium of $\overset{o}{E}$ is a triplet (f^*, y^*, p^*) with $f^* \in L_X$, $y^* \in Y_H$, $p^* \in R^n$, with $p^* > 0$, such that:

(i) a.e. in H:

$f^*(h)$ is maximal element for $\preccurlyeq_h$ of

$\gamma(h,p) = \{x \in X(h) \mid p^*x \leq p^*\overline{x}(h) + \pi(h,p^*)\}$;

(ii) $p^*y^* = \max p^*Y_H$;

(iii) $\int_H f^* \, d\mu = y^* + \int_H \overline{x} \, d\mu$.

For consumption possibility sets the same assumptions as for E are supposed to hold. Additionally, it is assumed that the following conditions are satisfied by the production sector of $\overset{o}{E}$: <u>B1</u>: $\exists\, Y$ such that $Y_C = \int_C Y \, d\mu$; <u>B2</u>: $Y_C \cap \Omega = \{0\}$ (impossibility of free production).

Finally, similar adequacy assumptions as for E are postulated to hold also for $\overset{o}{E}$.

(c) The Existence of Equilibria

It is not difficult to prove that if an equilibrium can be shown to exist for $\overset{o}{E}$, this includes the proof that an equilibrium exists also for E.

The strategy to prove that $\overset{o}{E}$ has an equilibrium is described next.

<u>Existence Theorem</u>. If $\overset{o}{E}$ satisfies all postulated assumptions, it has a quasi-equilibrium. Using the strong adequacy assumption, this quasi-equilibrium is an equilibrium.

A sequence $(\overset{o}{E}{}^k \mid k \in N)$ of truncated coalition production economies is defined such that each $\overset{o}{E}{}^k$ has bounded production and consumption sets, whose diameters increase with increasing k. The conventional analysis (see Section 2. on Finite Economies) is then applied to $\overset{o}{E}{}^k$.

Consider the equilibria sequence $((f^k, y^k, p^k) \mid k \in N)$. In general, $(f^k \mid k \in N)$ is not convergent in any standard topology. However, under suitable side conditions (Schmeidler (1970)), it has at least one adherent point in the a.e. pointwise convergence. This f^* together with the limits y^* and p^* makes a quasi-equilibrium of $\overset{o}{E}$ and so of E.

4. Economies with an Infinite-Dimensional Commodity Space

An infinite-dimensional commodity space appears, for instance, if any of the three characteristics identifying a commodity (physical properties, time, location) is allowed to vary continuously. Or (as supposed in this paper for illustration purposes), if the economic agents face an "infinite time horizon".

Work on this problem has been done by Nikaido (1956), Peleg and Yaari (1968), and Bewley (1972). In our current exposition we describe the model of the latter author.

(a) Basic Concepts and Definitions

The third model to be described consists of a finite number of households $(h = 1, \ldots, H)$, a finite number of firms $(f = 1, \ldots, F)$ and an infinite list of commodities

$$C = \bigcup_{i=1}^{\infty} \Gamma_t ,$$

where Γ_t is the finite set of commodities available at a time t (infinite time-horizon interpretation).

If $\mathcal{C}$ is the collection of all subsets of C, and μ a counting measure, we interpret $(C,\mathcal{C},\mu)$ as a σ-finite measure space. Our commodity space is now $L_\infty = L_\infty(C,\mathcal{C},\mu)$, the space of essentially bounded, real-valued, measurable functions on $(C,\mathcal{C},\mu)$.

L_∞ is endowed with the Mackey Topology. Let $ba = ba(C,\mathcal{C},\mu)$ stand for the set of bounded additive set functions on $(C,\mathcal{C})$, absolutely continuous in relation to μ, and $L_1 = L_1(C,\mathcal{C},\mu)$ for the set of countable additive set functions in ba $(L_1 \subset ba)$. Given the duality $\langle L_1, L_\infty \rangle$, $\sigma(L_1,L_\infty)$ represents the weakest topology on L_1 such that $L_1^* = L_\infty$ (topology of pointwise convergence on L). Similarly, $\sigma(L_\infty,L_1)$

represents the weakest topology on L_∞ such that $L_\infty^* = L_1$ (topology of pointwise convergence on L_1).

Consider again the duality $< L_1, L_\infty >$: $\tau(L_1, L_\infty)$ is the strongest topology on L_1 such that $L_1^* = L_\infty$, and $\tau(L_\infty, L_1)$ is the strongest topology on L_∞ for which $L_\infty^* = L_1$. They are called Mackey topologies, and are the topologies of uniform convergence on the $\sigma(L_\infty, L_1)$ - compact, convex, circled subsets of L_∞, and on the $\sigma(L_1, L_\infty)$ - compact, convex, circled subsets of L_1, respectively.

If L_∞ makes sense as the commodity space, L_1, its topological dual under the Mackey topology, makes sense, too, as the set of prices. However, the problem is not without complications. A price system may be regarded as a measurable function on $(C, \mathcal{C})$. The value of a commodity bundle (compare with Sections 2 and 3) is now given by

$$px = \int_C x(c)\, p(c)\, d\mu ,$$

where p is taken to be μ-integrable. L_1 is the set of all μ-integrable functions on C and every $p \in L_1$ is now a functional on L_∞ continuous with respect to the norm topology. However, under the norm topology $L_\infty^* = ba$, and for $p \in ba$, $p \notin L_1$ no reasonable economic interpretation exists. The need to restrict the analysis to L_1 ba, and how to do it, will be commented upon later.

A private ownership economy with a Mackey commodity space $L_\infty = L_\infty(C, \mathcal{C}, \mu)$ is a system

$$E = (X_h, \preccurlyeq_h, \overline{x}_h, Y_f, \alpha_{hf})$$

with $h = 1, \ldots, H$, and $f = 1, \ldots, F$. The consumption possibility set of h is

$$X_h \subset L_\infty^+ = \{x \in L_\infty \mid x \geqq 0\},$$

where $x \geqq 0$ if $x(c) \geqq 0$ a.e. It is totally preordered by a preference-indifference relation $\preccurlyeq_h$. Initial resources of h are $\overline{x}_h$.

Y_f is the production possibility set of the fth firm. Shares and profit distribution are described by the numbers $\alpha_{hf} \geqq 0$, with

$$\sum_{h=1}^{H} \alpha_{hf} = 1 \ , \ \forall f \ .$$

The budget set of h is still written

$$\{x \in X_h \mid px \leqq p\bar{x}_h + \sum_{f=1}^{F} \alpha_{hf} \cdot py_f\} \ .$$

The set of attainable states of the economy is

$$A = \{(x,y) \mid x_h \in X_h, \ y_f \in Y_f, \ \forall h \ \text{and} \ \forall f \ ; \ \text{and} \ x - y = \bar{x}\} \ .$$

An equilibrium of E is a triplet (x^*, y^*, p^*) such that

(i) $\forall h$, x_h^* is maximal element of the budget set of h for $\preccurlyeq_h$;

(ii) $\forall f$, $p^* \, y_f^* = \sup p^* \, Y_f$;

(iii) $x^* - y^* = \bar{x}$.

Assumptions on consumption sets are: <u>A1</u>: $\forall h$, X_h is a τ-closed convex subset of L_∞^+ ; <u>A2</u>: $\forall h$, $\preccurlyeq_h$ is a total preordering such that (a) $\forall x^o \in X_h$, $\{x \in X_h \mid x^o \preccurlyeq_h x\}$ is convex and -closed , and (b) $\forall x^o \in X_h$, $\{x \in X_h \mid x \preccurlyeq_h x^o\}$ is closed in the norm topology (note: we write τ for the Mackey topology on L_∞) .

Assumptions on the production sector of E are: <u>A3</u>: $\forall f$, Y_f is τ-closed, convex, and contains 0. Also: <u>A4</u>: <u>Monotonicity</u> - Consider in C subset E and define $K_E = \{x \in L_\infty \mid x(c) > 0 \ \text{a.e.} \ c \in E, \ x(c) = 0$ a.e. $c \notin E\}$. Then (assumption): $\exists \, C_c$ and $\exists \, C_p$, measurable sets of a partition $\{C_c, C_p\}$ of C such that:

(i) $\forall h$, if $x \in X_h$, and $k \in \tilde{K}_{C_c} = \{x \in K_{C_c} \mid \exists \, r > 0$ such that $x(c) \geqq r$ a.e. $c \in C_c\}$, then $x \preccurlyeq_h x + k$;

(ii) $\sum_{f=1}^{F} Y_f - K_{C_p} \subset \sum_{f=1}^{F} Y_f$;

(iii) $\mu(C_c) > 0$.

<u>A5</u>: <u>Adequacy</u> - $\forall h$, $\exists \, \bar{\bar{x}}_h \in X_h$ and $\exists \, \bar{\bar{y}}_h \in A(\sum_{f=1}^{F} Y_f)$

(here $A(\sum_{f=1}^{F} Y_f)$ is the asymptotic cone of $\sum_{f=1}^{F} Y_f)$, such that:

$\bar{\bar{y}}_h + \bar{x}_h - \bar{\bar{x}}_h >>> 0$ (here we write $x >>> 0$ if $\exists r > 0$ such that $x(c) \geq r$ a.e.). A6: Boundedness - f' e {1, ... , F} and any $u \in L_\infty$:

$$Y_{f'} \cap (L_\infty^+ - \sum_{\substack{f=1 \\ f \neq f'}}^{F} Y_f + u) \text{ is bounded}$$

(here a set Z is bounded if $\sup \{\|z\| \mid z \in Z\} < \infty$) .

(b) Equilibria with prices in $ba\ (C, \mathcal{C}, \mu)$

The existence theorem can be formulated as follows:

Existence Theorem. A private ownership economy E for which A1 - A6 hold has an equilibrium with price $p^* \in ba$.

The (Bewley's) method of proof is briefly described next. Define a sequence of finite economies (each E^k has a commodity space which is a finite-dimensional subspace of L_∞). It can be shown, by conventional methods, (Section 2), that each E^k has an equilibrium. It can also be shown that the set A of attainable states of E is bounded.

The spaces L_1 and L_∞ represent the equivalence classes, respectively, of $L_1 = L_1(C, \mathcal{C}, \mu) = \{f: C \to R \mid f$ is measurable, $\int |f| d\mu < \infty\}$ and

$$L_\infty = L_\infty(C, \mathcal{C}, \mu) = \{f: C \to R \mid f \text{ is measurable and } \|f\|_\infty < \infty\},$$

$$\text{where } \|f\|_\infty = \sup \{r \geq 0 \mid \mu\{c \mid |f(c)| \geq r\} > 0\} .$$

And, as is known, bounded subsets of L_∞ and of ba are $\sigma(L_\infty, L_1)$ and $\sigma(ba, L_\infty)$ relatively compact (every net has a convergent subnet). It can then be proved that the sequence of equilibria

$$((x^k, y^k, p^k) \mid k \in N)$$

converges to (x^*, y^*, p^*), which is an equilibrium of E with p^* in the set $ba(C,\mathcal{C},\mu)$.

(c) Restricting Prices to L_1

It was said in (a) that a price system $p \in ba$, $p \notin L_1$, does not make economic sense. Conditions under which equilibrium prices will be chosen from L_1 have been studied by Bewley (1972). Besides relatively minor modifications in the system A1 - A6 (which we shall not comment upon here) Bewley makes use of an "exclusion assumption" which allows him, by means of the Yosida-Hewitt Decomposition Theorems (1956), to separate within the equilibrium price system $p^* \in ba$ the countably additive part p_c^* from the purely finitely additive part p_p^*, $(p^* = p_c^* + p_p^*)$. By showing that p_p^* is concentrated on an arbitrarily small set of commodities (which are then excluded from E), it is proved that if (x^*, y^*, p^*) is an equilibrium with $p^* \in ba$, (x^*, y^*, p_c^*) is an equilibrium of the reduced economy with the relevant price system in L_1.

The suggested "exclusion assumption" can be formulated as follows. Let $\{F_n\}$ represent a sequence of measurable subsets of C, and p_c and p_p the countably additive and purely finitely additive parts of p, respectively; let χ_E denote the characteristic function of E, and Y a production possibility set. Then: B1: Exclusion - If $Y \subset L_\infty(C,\mathcal{C},\mu)$, then for any $p \in ba$, $\exists \{F_n\}$ such that (a) $\lim_n p_c(F_n) = 0$, $p_p(C \setminus F_n) = 0$, $\forall n$, and (b) if $y \in Y$, then $y\chi_{C \setminus F_n} \in Y$.

5. Perspectives

The three models described in this paper were all constructed with the same objective in mind. They all refer to a perfectly competitive system, and differ only in relation to two properties (one at a time) of the economy they are supposed to depict. Nevertheless, the solution of the same problem in the three cases did require fundamentally different topological "set ups".

Theoretical research is making its first serious steps in attempting to solve the equilibrium-existence problem under more general and realistic conditions. Rigorous models which try to make possible the integration of money, allow for indivisibilities and other nonconvexities, or attempt the description of an imperfectly competitive reality, already suggest the need for new "set ups" of an increasing complexity. In these areas, however, research is still at its beginnings.

6. Mathematical Sources

The elementary topological concepts of Section 2 are discussed in Klein (1973). See also Nikaido (1968).

For the mathematics of Section 3, the books by Neveu (1965), and by Dunford and Schwartz (1966) can be consulted.

The tools of Section 4 are found in Dunford and Schwartz (1966), Kelley and Namioka (1963), Schaeffer (1966), and Treves (1967).

References

[1] Arrow, K.H. and Hahn, F.H.: "General Competitive Analysis", Holden Day, San Francisco, Edinburgh (1971).

[2] Aumann, R.: "Markets with a continuum of traders", Econometrica 32 (1964), 39 - 50.

[3] __________: "Existence of equilibria in markets with a continuum of traders", Econometrica 34 (1966), 1 - 17.

[4] __________: "Disadvantageous monopolies", Journal of Economic Theory 6 (1973), 1 - 11.

[5] Bewley, T.: "Existence of equilibria in economies with infinitely many commodities", Journal of Economic Theory 4 (1972), 514 - 540.

[6] Debreu, G.: "Theory of Value: An Axiomatic Analysis of Economic Equilibrium", John Wiley, New York (1959).

[7] Debreu, G. and Schmeidler, D.: "The Radon-Nykodym derivative of a correspondence", CORE (1970).

[8] Dunford, N. and Schwartz, J.: "Linear Operators, Part I", Interscience, New York (1966).

[9] Hildenbrand, W.: "On a core of an economy with a measure space of agents", Review of Economic Studies 35 (1968), 443 - 452.

[10] __________: "Existence of equilibria for economies with production and a measure space of consumers", Econometrica 38 (1970), 608 - 623.

[11] Kannai, Y.: "Continuity properties of the core of a market" (revised version), Research Memorandum 34 Department of Mathematics, The Hebrew University, Jerusalem (1968).

[12] Kelley, J.L. and Namioka, I.: "Linear Topological Spaces", Van Nostrand, New York (1963).

[13] Klein, E.: "Mathematical Methods in Theoretical Economics: Topological and Vector Space Foundations of Equilibrium Analysis", Academic Press, New York, London (1973).

[14] Neveu, J.: "Mathematical Foundations of the Calculus of Probability", Holden Day, San Francisco (1965).

[15] Nikaido, H.: "Convex Structures and Economic Theory", Academic Press, New York, London (1968).

[16] __________: "On the existence of competitive equilibrium for infinitely many commodities", Technical Report No. 34, Stanford University, mimeographed (1956).

[17] Pareto, W.: "Manuale di Economia Politica", Societa Editrice Libraria, Milano (1906).

[18] Peleg, B. and Yaari, M.: "Markets with countably many commodities", Research Memorandum 37 Department of Mathematics, The Hebrew University, Jerusalem (1968).

[19] Schaeffer, H.: "Topological Vector Spaces", Macmillan, New York (1966).

[20] Schmeidler, D.: "Competitive equilibria in markets with a continuum of traders and incomplete preferences", Econometrica 37 (1969), 578 - 585.

[21] Schmeidler, D.: "Fatou's lemma in several dimensions", Proc. Amer. Math. Soc. 24 (1970), 300 - 306.

[22] Treves, F.: "Topological Vector Spaces, Distributions and Kernels", Academic Press, New York, London (1967).

[23] Vind, K.: "Edgeworth allocations in an exchange economy with many traders", International Economic Review 5 (1964), 165 - 177.

[24] Walras, L.: "Elements d'economie politique pure", Corbaz, Lausanne (1874-1877).

[25] Yosida, K. and Hewitt, E.: "Finite additive measures", Trans. Amer. Math. Soc. 72 (1956), 46 - 66.

EQUIVARIANT EXTENSIONS OF MAPS FOR COMPACT LIE GROUP ACTIONS

Jan W. Jaworowski

Indiana University

The paper studies extension and retraction properties in the category Top^G of spaces with compact Lie group actions. The existence of an extension of an equivariant map does not, in general, imply the existence of an equivariant extension. However, the following extension theorem is proved:

If a compact Lie group G acts on a finite dimensional compact metric space X with a finite number of orbit types, if $\tilde{X}$ is a closed equivariant subspace of X containing all the fixed points (so that the action on $X - \tilde{X}$ is free) and if $f: \tilde{X} \to Y$ is an equivariant map to a compact metric space Y with a G-action, then an equivariant neighbourhood extension of f exists, provided that Y is an ANR; and if Y is an AR, then f can be extended over the whole of X.

This theorem is then applied to a characterization of absolute retracts in the category Top^G. At least in the abelian case, and provided that the number of isotropy subgroups is finite, it is shown that a G-space is an equivariant ANR (i.e., a G-ANR) if and only if the fixed point set of every subgroup of G is a (topological) ANR. A similar condition characterizes G-AR's.

PROPERTIES OF FUNCTIONS WITH A CLOSED GRAPH

Ivan Baggs

St. Francis Xavier University, Antigonish

1. Introduction. Let X and Y be topological spaces and let f be a function from X into Y. The function f has a closed graph if $G(f) = \{(x,f(x)) \mid x \in X\}$ is a closed subset of $X \times Y$. Put $D(f) = \{x \in X \mid f \text{ is discontinuous at } x\}$. (Throughout, all spaces are assumed to be at least Hausdorff.)

In this note we will be concerned with two problems involving functions with a closed graph. The first of these is related to a well-known characterization of the points of discontinuity of an arbitrary real valued function of a real variable. Namely, a set $F \subset R$ is an F_σ set if and only if there exists a function $f : R \to R$ such that $D(f) = F$ (see [10], p. 171). In section 3, we investigate the set of points of discontinuity of a real valued function with a closed graph. The second problem which will concern us, will be to find conditions under which a function with a closed graph is continuous (see section 4).

2. Preliminaries. In this section we present some of the known properties of functions with a closed graph which are used in proving the results in sections 3 and 4.

Property 1. Let f be a function from X into Y. $D(f)$ is closed if and only if for each net x_α in X, where $x_\alpha \to x$ and $f(x)_\alpha \to y$, it follows that $f(x) = y$ (see [9], p. 195).

Property 2. Let $f : X \to Y$ be a function with a closed graph. If K is a compact subset of Y, then $f^{-1}(K)$ is a closed subset of X. ([7])

Property 3. Let $f : X \to Y$ be a function with a closed graph. If K is a compact subset of X, then $f(K)$ is a closed subset of Y (see [9], p. 196).

Property 4. Let f be a function from X into Y, where Y is compact. $G(f)$ is closed if and only if f is continuous ([3]).

Property 5. Let $f : X \to Y$ be any function, where Y is locally compact. If for each compact set $K \subset Y$, $f^{-1}(K)$ is closed, then $G(f)$ is closed ([6]).

3. Points of discontinuity. It is known (see [5], p. 78) that in order for a subset F of a topological space to coincide with the set of points of discontinuity of a real valued function on that space, it is necessary that F be an F_σ set without isolated points. It has recently been shown by Bolstein ([2]) that this condition is also sufficient for a wide class of topological spaces. In particular, it is sufficient if the space is separable, first countable, locally compact or a linear topological space.

Functions with a closed graph have several properties which are, in many ways, similar to properties of continuous functions. (See, for example, Properties 1,2, and 3.) Instead of considering arbitrary functions, as above, we asked the following question: if a real valued function has a closed graph, what can be said about the set of points where the function is discontinuous? The following results are given in [1].

Theorem 1. Let X be a Baire space. If $f : X \to R$ has a closed graph, then $D(f)$ is a closed and nowhere dense subset of X.

Theorem 2. A set $F \subset R$ is closed and nowhere dense if and only if there exists a function $f : R \to R$ such that f has a closed graph and $D(f) = F$.

There exists a Baire space X and a closed nowhere dense subset F of X such that F is not the set of points of discontinuity of a real valued function on X with a closed graph. For example, let X be the space of all ordinals less than or equal to the first uncountable ordinal, Ω, with the order topology. Put $F = \{\Omega\}$. Let $f : X \to R$ be any function with a closed graph. $X - F$ is pseudocompact, so, if f is continuous on $X - F$, f is bounded on $X - F$. This implies that $D(f) \neq F$.

4. In this section we investigate some conditions under which functions with a closed graph become continuous. First we recall the following definitions:

Definition 1. A function $f : X \to Y$ is connected if, for each connected subset C of X, $f(C)$ is a connected subset of Y.

Definition 2. A function $f : X \to Y$ is sequentially closed if, for each point $x \in X$ and for each sequence $x_n \to x$, $\{f(x_n) \cup f(x)\}$ is a closed subset of Y.

It is easily seen that if a function $f : X \to Y$ has a closed graph, then f is sequentially closed. However, a sequentially closed function may not have a closed graph. For example, define $f : R \to R$ as follows: $f(x) = 1$, if $x \neq 0$, and $f(0) = 0$. f is sequentially closed but f does not have a closed graph.

Theorem 3. Let X be a first countable and locally connected topological space. If $f : X \to R$ is almost closed and connected, then f is continuous.

Proof. Suppose f is not continuous at $x_0 \in X$. Then there exists a sequence $x_n \to x_0$ such that $f(x_n)$ does not converge to $f(x_0)$. Let U_m, $m = 1,2,\ldots,$ be a neighbourhood base at x_0 such that, for each m, U_m is connected.

If there exists a subsequence x_k, $k = 1,2,\ldots,$ of x_n, $n = 1,2,\ldots,$ such that $\{f(x_k) \cup f(x_0)\}$ has no limit point in R, then $\{f(x_k) \cup f(x_0)\}$ is unbounded. This implies that $f(U_m)$ is unbounded for each m. Since f is a connected function, we may assume without loss of generality, that $f(U_m) \supset [f(x_0),+\infty)$, for $m = 1,2,\ldots$. Let $q \in (f(x_0),+\infty)$. Select a sequence of distinct points q_m , such that $q_m \in f(U_m)$, for each m, $q_m \to q$ and $q_m \neq q$ for all m. Let

$$q_m' \in [f^{-1}(q_m)] \cap U_m ,$$

for $m = 1,2,\ldots$. Then $q_m' \to x_0$. It follows that q is a limit point of $\{f(q_m') \cup f(x_0)\}$ but is not a point of that set. Since f is sequentially closed, this is impossible. Therefore, $\{f(x_k) \cup f(x_0)\}$ is bounded.

Since f is not continuous at x_0 , it now follows that there must exist some $p \in R$ such that $p \neq f(x_0)$ and p is a limit point of $\{f(x_n) \cup f(x_0)\}$. As before, select a sequence of distinct points p_m , such that $p_m \in f(U_m)$, for each m , $p_m \to p$ and $p_m \neq p$, for all m (this is possible since $f(U_m)$ is connected). Let

$$p_m' \in [f^{-1}(p_m)] \cup U_m .$$

Then $\{f(p_m') \cup f(x_0)\}$ is not a closed set, which contradicts the fact that f is sequentially closed. Therefore, f must be continuous at x_0 and the theorem is established.

Corollary 1. Let X be a first countable and locally connected topological space. If $f : X \to R$ is a connected function with a closed graph, then f is continuous.

Corollary 2. Let $F : R \to R$ be a function. If f is the derivative of F(x) and if f has a closed graph, then f is continuous.

Remark 1. If $f : R \to R$ is a function with a closed graph and if f is not continuous at x_0 , then f is not a connected function. In which case, there is an interval I containing x_0 such that if I' is a subinterval of I containing x_0 , then f(I') is not connected.

The following definition is given by Hamilton ([4]).

Definition 3. A function $f : X \to Y$ is peripherally continuous if for each $x \in X$ and for each pair of open neighbourhoods U and V containing x and $f(x)$, respectively, there exists an open set $G \subset U$ and containing x such that f maps the boundary of G into V.

A peripherally continuous function may be discontinuous everywhere. For example, define $f : R \to R$ by $f(x) = 1$, if x is rational, and $f(x) = 0$, otherwise.

Theorem 4. Let $f : R \to R$ be a function with a closed graph. If f is peripherally continuous, then f is continuous.

Proof. Suppose f is not continuous at some point $x \in R$. Then, by Remark 1, we may assume that there exists some $b > x$ such that if $I = [x,b]$, $f(I)$ is not connected. Therefore, $f(I) \subset A_0 \cup A_1$, where A_0 and A_1 are two disjoint intervals. Since I is compact, it follows from Property 3 that $f(I)$ is closed.

If A_0 is bounded, then, since $f(I)$ is closed, there must exist a closed and bounded set $F_0 \subset A_0$, such that for each $y \in I$, where $f(y) \in A_0$, $f(y) \in F_0$. Since F_0 is a closed and bounded subset of R and since $f|I$ also has a closed graph, Property 2 implies that $(f|I)^{-1}(F_0)$ is a closed subset of I. Therefore

$$\{(f|I)^{-1}(F_0)\}^c = \bigcup_{n=1}^{\infty} G_n ,$$

where $G_n = (a_n,b_n)$, for $n = 1,2,\ldots$. Since $f(I)$ is not connected,

$$\{(f|I)^{-1}(A_1)\} \cap G_n \neq \phi , \text{ for some } n .$$

Without loss of generality we may assume that $G_1 \subset (f|I)^{-1}(A_1)$. But since $f(a_1)$ and $f(b_1)$ are elements of A_0, this contradicts the fact that f is peripherally continuous. Hence A_0 cannot be bounded. Similarly, A_1 is not bounded.

Since A_0 and A_1 are arbitrary open sets which separate $f(I)$, it follows that if A and B are any two sets which separate $f(I)$, then both A and B must be unbounded. Therefore, if $f(x) \in A_0$, this implies that A_0 contains an unbounded interval containing $f(x)$. It now follows from Remark 1 that if $x \leq a \leq b$, then $f([x,a])$ contains an unbounded interval containing $f(x)$. This contradicts the fact that f has a closed graph. Therefore, f must be continuous at x and the theorem is established.

It was proven by W. Sierpinski that a connected and compact metric space cannot be written as a countable union of disjoint closed sets. It has also been shown that if X is any compact and connected Hausdorff space then X cannot be written as a countable union of disjoint closed sets (see [9], p. 209).

Theorem 5. Let X be a compact and connected Hausdorff space and let Y be Hausdorff. If there exists a non-constant function f with a closed graph from X onto Y, then Y is uncountable.

Proof. Suppose Y is countable. Let $\{y_k \mid k = 1,2,\ldots\}$ be an enumeration of Y. Since f has a closed graph, $f^{-1}(y_k)$ is a closed subset of X, for each k, and X can be written as a countable union of disjoint closed sets. Since this is impossible by the preceeding remarks, the theorem is established.

References

[1] Baggs, I., Functions with a closed graph. To appear in Proc. Amer. Math. Soc.

[2] Bolstein, R., Sets of points of discontinuity. Proc. Amer. Math. Soc. 38 (1973), 193-197.

[3] Dugundji, J., Topology. Allen and Bacon, Inc. Boston, 1966.

[4] Hamilton, O. H., Fixed points for certain non-continuous transformations. Proc. Amer. Math. Soc. 8 (1957), 750-756.

[5] Hewitt, E. and Stromberg, K., Real and Abstract Analysis. Springer-Verlag, New York, 1965.

[6] Long, P. and McGehee, E. E., Properties of almost continuous functions. Proc. Amer. Math. Soc. (1970), 175-180.

[7] Fuller, R. V., Relations among continuous and various non-continuous functions. Pacific J. Math. 25 (1968), 495-509.

[8] Sprecher, D. A., Elements of Real Analysis. Academic Press New York, 1970.

[9] Wilansky, A., Functional Analysis. Alaisdell Publ. Co., New York, 1964.

[10] Willard, S. W., General Topology. Addison Wesley, Reading, Mass. 1970.

A UNIFIED APPROACH TO SOME BASIC PROBLEMS IN HOMOTOPY THEORY

Peter Booth

Memorial University, St. John's

It will be shown that many problems in homotopy theory are equivalent to problems involving the restricted fibered mapping projection (p q;a). Techniques for the solution of these equivalent problems will be discussed; in particular the exact cohomology sequences of Serre, Wang and Gysin will be derived (cf. P.I. Booth, A unified treatment of some basic problems in homotopy theory, Bull. Amer. Math. Soc. 79 (1973), 331-336).

STABLE COMPLEX STRUCTURES ON REAL MANIFOLDS

Allan Brender

University of Arizona

We discuss a weakened version of the classical problem of integrating an almost complex structure on a real manifold. It is shown that if M admits an a.c. structure then the product of M and some Euclidean space is a complex manifold, in fact, a Stein manifold. Such Stein structures are then classified, up to deformation with respect to a real parameter, by reductions of the stable normal bundle of M to complex bundles. We also discuss the relation of these results to work of Landweber on Haefliger's classifying space for complex structures.

NONEXISTENCE OF AXIAL MAPS

Donald M. Davis

Northwestern University, Evanston

1. Discussion of Results

An axial map is a map of real projective spaces $P^i \times P^j \to P^n$ whose restriction to $P^i \times * \cup * \times P^j$ is the inclusion map. These were first defined by Hopf [3] and have also been studied by James [4] and Gitler. Their importance arises from their relation to the geometric dimension of vector bundles over P^j and hence to immersions of projective space in Euclidean space:

Proposition [3] If the n-fold Whitney sum of the canonical line bundle over P^j, $n\xi_j$, has an i-dimensional trivial subbundle, then there is an axial map $P^{i-1} \times P^j \to P^{n-1}$.

We shall use connective k-theory and Adams operations to detect nonexistence of some axial maps.

Theorem If $\binom{n}{i}$ is odd, then there does not exist an axial map $P^i \times P^{n-i+2\nu(n+1)+2} \to P^n$, where $\nu(2^a(2b+1)) = a$.

The +2 can sometimes be decreased by as much as 3 depending upon mod 4 values of i and $\nu(n+1)$. For precise results see [2], where this theorem and another are proved by slightly different techniques. The advantage of the method used herein is that it makes it easier to consider what happens when $\binom{n}{i}$ is even. The theorem becomes interesting only when $n+1$ is a multiple of 32. We then obtain many new results such as geometric dimension of $32\xi_i \geq i-8$ for $i = 10, 14, 26, 30$.

2. A Sketch of the Technique

We first use the Atiyah-Hirzebruch spectral sequence to compute $\tilde{k}_u^*(P^n)$ and $\tilde{k}_u^*(P^i \wedge P^j)$. The coefficient groups $\pi_i(b_u)$ are Z in nonnegative even dimensions, so that

$$E_2^{p,q}(P^n) = \begin{cases} Z_2 & \text{if } p,q \text{ even, } 0 \le p \le n,\ 0 \le q \\ Z & \text{if } n \text{ odd, } p=n,\ q \text{ even, } 0 \le q \\ 0 & \text{otherwise} \end{cases}$$

which we depict (for odd n) by

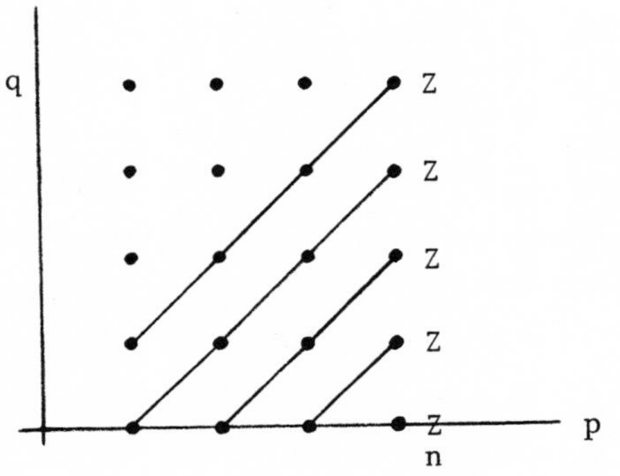

There are no differentials by dimensional considerations and the extensions are nontrivial by [1], as indicated by the slanting lines, so that the even dimensional groups are cyclic of order a power of 2, and if n is odd, the odd dimensional groups are infinite cyclic.

There are d_2 - differentials in the spectral sequence for $P^i \wedge P^j$, due to the action of Sq^3 in $H^*(P^i \wedge P^j; Z)$. When i and j are even, $E_3^{p,q}(P^i \wedge P^j)$ can be depicted, except for some Z_2 classes in $E_3^{p,0}$, by

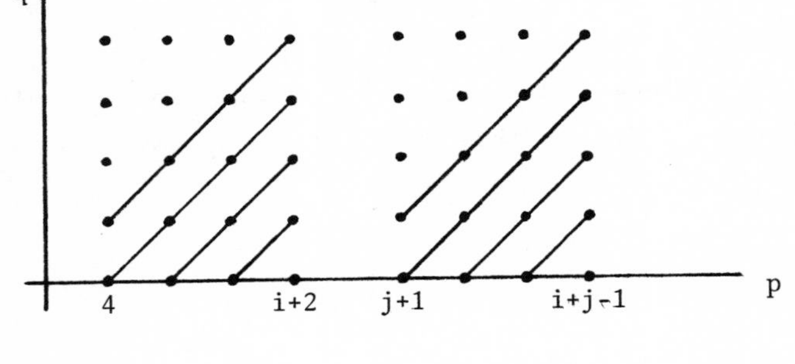

and $E_3 = E_\infty$.

If $\alpha : P^i \times P^j \to P^n$ is an axial map, and g is the generator in $H^1(P;Z_2)$, then $\alpha^*(g) = g_1 \otimes 1 + 1 \otimes g_2$ and hence $\alpha^*(g^b) = \Sigma\binom{b}{a} g_1^a \otimes g_2^{b-a}$. The Hopf construction yields a map $\Sigma(P^i \wedge P^j) \xrightarrow{\overline{\alpha}} \Sigma P^n$ whose cohomology is similar to that of α. We prefer $P^i \wedge P^j$ to $P^i \times P^j$ because it has fewer extraneous k-theory classes. This determines the map of Atiyah-Hirzebruch spectral sequences and hence determines the map in k_u-theory up to elements of higher filtration. Indeed, if i and j are even and n is odd, $\alpha^* : Z \approx k_u^n(P^n) \to k_u^n(P^i \wedge P^j) \approx Z/2^{(i+j-n+1)/2}$ sends generator to generator if and only if $\binom{n}{i}$ is odd. We shall later consider how to determine α^* when $\binom{n}{i}$ is even.

If we localize at 2, b_u and b_o admit Adams operations ψ^3 which can be computed from [1] to be multiplication by $3^{(n+1)/2}$ in $k_u^n(P^i \wedge P^j)$ and the identity in $k_u^n(P^n)$. Thus if $\binom{n}{i}$ is odd, then

$$3^{\frac{n+1}{2}} G_2 = \psi^3\alpha^* G_1 = \alpha^*\psi^3 G_1 = G_2 ,$$

and hence $(3^{\frac{n+1}{2}} - 1)\, G_2 = 0$ in $Z/2^{(i+j-n+1)/2}$. Thus if $\frac{i+j-n+1}{2} > \nu(3^{(n+1)/2} - 1) = \nu(2(n+1))$, there can be no axial map.

A similar program can be carried out using k_o^* instead of k_u^* with the advantage that $k_o^n(P^i \wedge P^j)$ is sometimes twice as large as $k_u^n(P^i \wedge P^j)$ so that stronger results are obtained. We can obtain even stronger results if we apply k_o^* to $P^i \wedge P^j \wedge D(P^n) \xrightarrow{\alpha\wedge 1} P^n \wedge D(P^n)$, where $D(P^n)$ indicates the Spanier-Whitehead dual P^{L-2}/P^{L-n-2}, with L a large power of 2. The groups and Adams operations can be computed with some difficulty as in [2;Section 4] and $(\alpha \wedge 1)^*$ can for the most part be computed from knowledge of α^*. In some cases $k_o^{L-1}(P^i \wedge P^j \wedge D(P^n))$ is twice as large as $k_o^n(P^i \wedge P^j)$, and the result obtained by applying ψ^3 to the map in k_o^{L-1} induced by α yields the theorem.

The applicability of the theorem could be greatly broadened by a statement about α^* when $\binom{n}{i}$ is even. The natural conjecture is that $\alpha^* G_1 = \binom{n}{i} G_2$. This is apparently not true, for it would imply

$$P^{24} \not\subseteq R^{43},$$

which contradicts a result of [5].

If $\binom{n}{i}$ is even, then $\alpha^*G_1 = 2G_2$ if and only if the integral functional cohomology operation $\widetilde{Sq}^3_\alpha(G_1)$ is nonzero (mod the image of $\widetilde{Sq}^3$). To see this, consider the diagram

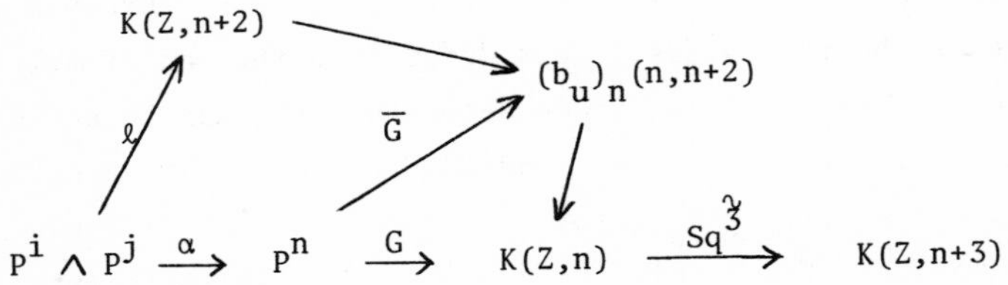

where $(b_u)_n(n,n+2)$ has as its only nonzero homotopy groups the nth and $(n+2)^{nd}$ groups of the nth space in the b_u-spectrum. Then $4k_u^n(P^i \wedge P^j) = \ker(k_u^n(P^i \wedge P^j) \to [P^i \wedge P^j, (b_u)_n(n,n+2)])$, so $\alpha^*(G_1)$ is a multiple of 4 if and only if $\bar{G}\alpha$ is trivial if and only if ℓ is trivial. But this is precisely $\widetilde{Sq}^3_\alpha(G_1)$. I have been unable to evaluate this operation.

References

[1] J. F. Adams, "Vector fields on spheres", Ann. of Math. 75 (1962), 603-632.

[2] D. M. Davis, "Generalized homology and the generalized vector field problem", to appear in Quart. J. Math. Oxford.

[3] H. Hopf, "Ein topologischer Beitrag zur reellen Algebra", Comm. Math. Helv. 13 (1941), 219-239.

[4] I. M. James, "On the immersion problem for real projective spaces", Bull. Amer. Math. Soc. 69 (1963), 231-238.

[5] A. D. Randall, "Some immersion theorems for projective spaces", Trans. Amer. Math. Soc. 147 (1970), 135-151.

EQUIVALENCE RELATIONS AND TOPOLOGY

A.E. Fekete

Memorial University, St. John's

In the lattice QX of all quotient sets of X the order relation is defined such that for $A, B \in QX$, $A \sqsubseteq B \Leftrightarrow \alpha = \gamma\beta$, where $\alpha: X \to A$, $\beta: X \to B$ are the natural surjections. The least upper bound and greatest lower bound of $\{A_i\}$, $A_i \in QX$, are denoted by $\bigsqcup_i A_i$ and $\bigsqcap_i A_i$, resp., and will be called the superimposition and amalgamation of quotient sets.

Q-topological structure on a set X is introduced in terms of this lattice as follows: a family $\{A_i\}$, $A_i \in QX$ is distinguished by its elements being called closed quotient sets of X, such that

(i) $\not{1}$, X are closed ($\not{1}$ is the trivial quotient set)

(ii) A_1, A_2 closed $\Rightarrow$ $A_1 \sqcap A_2$ closed

(iii) A_j, $(j \in I)$, closed $\Rightarrow$ $\bigsqcup_j A_j$ closed.

Q-continuity of $f: X \to Y$ is defined by stipulating that the induced function $f^Q: QX \to QY$ sending $A \in QX$ into its pushout $f^Q(A) \in QY$ preserve closed quotient sets. Since Q is a covariant functor, the product of continuous functions is continuous. The closure $\overline{A}$ of an arbitrary quotient set $A \in QX$ is introduced as the coaxis of the fixgroup of Q-homeomorphisms of A. Then $\overline{A}$ is closed for every $A \in QX$. Convergent filters and ultrafilters are defined and some standard theorems proved.

AN EXAMPLE OF TOPOLOGICAL CONJUGACY IN NON-EQUILIBRIUM STATISTICAL MECHANICS

M. Grmela

Carleton University, Ottawa

It is a well-known experience that dynamics of many various fluids in the vicinity of single phase thermodynamically stable equilibrium states can be well approximated by hydrodynamic equations. The individuality of each particular fluid enters only in coefficients appearing in hydrodynamic equations. On the other hand, any microscopic dynamics of different fluids can differ substantially. This problem is mathematically formulated and discussed, by using ideas of topological dynamics, for an example of microscopic dynamics defined by a family of kinetic equations of the Enskog-Vlasov type.

CHANGE OF BASE POINT AND FIBRATIONS

Philip R. Heath

Memorial University, St. John's

It is well known that if x_0 is a nondegenerate base point of a topological space X, and (Y,y_0) is arbitrary, then for each path $\omega: y_0 \to y_1$ in Y there is a bijection $\omega_\# : [X,x_0 ; Y,y_0] \to [X,x_0 ; Y,y_1]$ of sets of based homotopy classes. Little consideration seems to have been given to change of base point in X. We give here a method of obtaining both results from standard constructions on fibre spaces.

Let X be a locally compact Hausdorff space and let Y be arbitrary. Since x_0 is a nondegenerate base point it follows that the map $x_0^* : Y^X \to Y$ defined by $x_0^*(f) = f(x_0)$ is a fibration. The fibre of $y_0 \in Y$ over x_0^* is the subspace $Y^X(x_0,y_0)$ of Y^X consisting of maps from X to Y which take x_0 to y_0. It is clear that $\pi_0(Y^X(x_0,y_0)) = [X,x_0 ; Y,y_0]$.

Let $p : E \to B$ be a fibration and let $H : f_0 \simeq f_1 : Z \times I \to B$ be a homotopy. It is well known (see for example §7 of [3]) that H induces a homotopy equivalence $H^*E : f_0^*E \to f_1^*E$ where for example f_0^*E denotes the pullback of f_0 and p. In the case where Z is a one point space, $B = Y$, and p is the fibration x_0^* above, H becomes a path from y_0 to y_1 say, in Y, y_i^*E becomes $Y^X(x_0,y_i)$, $(i = 0,1)$, and H induces a homotopy equivalence $H^*Y^X : Y^X(x_0,y_0) \to Y^X(x_0,y_1)$. Applying the path component functor π_0 we obtain the standard result:

<u>Theorem 1</u>. Let X be a locally compact Hausdorff space with nondegenerate base point x_0 and let Y be arbitrary. Then for each path $\omega: y_0 \to y_1$ in Y there is a bijection

$$\omega_\# = \pi_0\omega^*(Y^X) : [X,x_0 ; Y,y_0] \to [X,x_0 ; Y,y_1].$$

Suppose now that $\lambda : x_0 \to x_1$ is a path in X between two non-degenerate base points x_0 and x_1. Clearly both x_0^* and $x_1^* : Y^X \to Y$ are fibrations. Moreover λ induces a homotopy $\lambda^! : x_0^* \simeq x_1^* : Y^X \times I \to Y$ (see for example [1, Proposition 3.8]). Moreover diagram 2a is commutative.

Diagram 2a
$$\begin{array}{ccc} Y^X & \xrightarrow{\ 1\ } & Y^X \\ \downarrow & & \downarrow \, x_o^* \\ Y^X \times I & \longrightarrow & Y \end{array}$$
$$\lambda^! : x_o^* \simeq x_1^*$$

Diagram 2b
$$\begin{array}{ccc} Y^X & \xrightarrow{\ \lambda_\delta\ } & Y^X \\ & {\scriptstyle x_1^*} \searrow \quad \swarrow {\scriptstyle x_o^*} & \\ & Y & \end{array}$$

(The left hand map is the inclusion of Y^X at $Y^X \times \{0\}$). Since x_0^* is a fibration, $\lambda^!$ lifts to a homotopy $\lambda^+ : 1 \simeq \lambda$ say, and λ_δ makes diagram 2b commutative. Also λ_δ, being homotopic to a homotopy equivalence $(1 : Y^X \to Y^X)$ is itself a homotopy equivalence. Let $\overline{\lambda}_\delta$ denote the restriction of λ_δ to the fibres over y_0; thus $\overline{\lambda}_\delta : Y^X(x_1,y_0) \to Y^X(x_0,y_0)$. Then $\overline{\lambda}_\delta$ is a homotopy equivalence by corollary 1.5 of [2], applying the functor π_0 we obtain a bijection $\pi_o\overline{\lambda}_\delta$. This bijection is independent of the choice of λ_δ, since if λ'_δ is the end point of any other lift of $\lambda^!$ then λ_δ and λ'_δ are homotopic rel x_0^*, and it follows that the restrictions $\overline{\lambda}_\delta$ and $\overline{\lambda}'_\delta$ are homotopic. Collecting this information together we have:

<u>Theorem 3.</u> Let $\lambda : x_0 \to x_1$ be a path between nondegenerate base points in a locally compact Hausdorff space X, and let (Y,y_0) be arbitrary. Then λ induces a bijection

$$\lambda_\# = \pi_o\overline{\lambda}_\delta : [X,x_1 \,;\, Y,y_0] \to [X,x_0 \,;\, Y,y_0].$$

The bijection of Theorem 1 is well known to be functorial, and $\omega_\#$ is the evaluation of $[\omega] \in \pi B$, the fundamental groupoid of B, under the composite

$$\pi B \xrightarrow{\ \pi^* x_0^*\ } \text{H Top} \xrightarrow{\ \pi_0\ } \text{Set},$$

where $\pi^* x_0^*$ is defined in Corollary 7.7 in [3]. Theorem 1.2 of [4] applied to the map $(\lambda_\delta, 1)$ of diagram 2b above shows the existence of a natural transformation $\Phi : \pi^* x_1^* \to \pi^* x_0^*$ defined on each point $y_0 \in \pi Y$ to be precisely cls $\overline{\lambda}_\delta$ where λ_δ is defined above. Corollary 1.3 of [4] shows

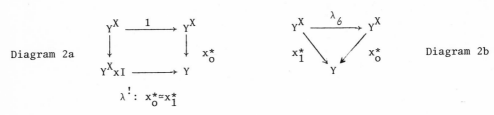

that in this case Φ is a natural equivalence. Once more composing with π_o we obtain:

Theorem 4. Under the conditions of Theorems 1 and 3 there is a commutative diagram of bijections

$$\begin{array}{ccc} [X,x_1 \,;\, Y,y_o] & \xrightarrow[\lambda_\#]{} & [X,x_o \,;\, Y,y_o] \\ \downarrow \omega_\# & & \downarrow \omega_\# \\ [X,x_1 \,;\, Y,y_1] & \xrightarrow[\lambda_\#]{} & [X,x_o \,;\, Y,y_1] \end{array}$$

The above results generalise to the relative case; suppose x_o is a nondegenerate base point for the pair (X,A) of spaces and that (Y,B) is an arbitrary pair, then the map $x_o^* : (Y,B)^{(X,A)} \to B$ is a fibration ($(Y,B)^{(X,A)}$ is the obvious subspace of Y^X). Moreover the path components of the fibre of x_o^* over y_o is the set $[X,A,x_o \,;\, Y,B,y_o]$.

Theorem 5. Let $\lambda : x_o \to x_1$ be a path (in A) between two nondegenerate base points of the pair (X,A) of locally compact Hausdorff spaces. Further let $\omega : y_o \to y_1$ be a path (in B) in the arbitrary pair (Y,B). Then there is a commutative diagram of bijections

$$\begin{array}{ccc} [X,A,x_1 \,;\, Y,B,y_o] & \xrightarrow[\lambda_\#]{} & [X,A,x_o \,;\, Y,B,y_o] \\ \downarrow \omega_\# & & \downarrow \omega_\# \\ [X,A,x_1 \,;\, Y,B,y_1] & \xrightarrow[\lambda_\#]{} & [X,A,x_o \,;\, Y,B,y_1] \end{array}$$

Bibliography

[1] R. Brown : Function spaces and product topologies. Quart. J. Math. Oxford (ser. 2) 15 (1964), 238-250.

[2] R. Brown and P.R. Heath : Coglueing homotopy equivalences. Math. Z. (1970), 313-325.

[3] P.R. Heath : Groupoid operations and fibre homotopy equivalence I. Math. Z. 130 (1973), 207-233.

[4] P.R. Heath : Groupoid operations and fibre homotopy equivalence II. (Manuscript).

K(X) AS A COMODULE

P. Hoffman

University of Waterloo

We state some results whose proofs will appear elsewhere [5], then indicate a few applications and problems. X denotes a finite complex.

Let $L(t)^*$ be the cohomology theory with coefficients in $\mathbb{Q}_p$, the integers localized at p, represented by the spectrum whose $2n^{th}$ term is $BU(2n,\ldots,2n+2t)$. The Chern character provides an isomorphism

$$L(t)^{2n}(X) \otimes \mathbb{Q} \cong \bigoplus_{i=0}^{t} H^{2n+2i}(X;\mathbb{Q}).$$

The splitting of the right side gives a set $\{\pi_0,\pi_1,\ldots,\pi_t\}$, of stable natural idempotent operations on $L(t)^* \otimes \mathbb{Q}$, which is a basis for a $\mathbb{Q}$-Hopf algebra $A(t)^* \otimes \mathbb{Q}$, whose dual is $\mathbb{Q}[\eta]/(\eta^{t+1})$, where $\langle\pi_i,\eta\rangle = \delta_{i1}$. Let

$$A(t)^* = \{\theta \in A(t)^* \otimes \mathbb{Q} \mid \theta[L(t)^*(X)] \subset L(t)^*(X) \text{ if } H^*(X;\mathbb{Q}_p) \text{ free},$$

$$A(t) = \{f \in \mathbb{Q}[\eta]/(\eta^{t+1}) \mid \langle\theta,f\rangle \in \mathbb{Q}_p \text{ for all } \theta \in A(t)^*\}.$$

<u>Theorem A.</u> $A(t) = \text{Span}_{\mathbb{Q}_p}\{f_{j_1}f_{j_2}\cdots f_{j_r}+(\eta^{t+1}) \mid \sum j_i \leq t\}$,

where
$$f_j = \eta^{-1}\binom{\eta}{j+1}.$$

To prove this, one shows that $\theta \in A(t)^*$ as long as $\theta[L(t)^*(X)] \subset L(t)^*(X)$, when X is a product of complex projective spaces. Then set up a coaction $\lambda: L(t)^*(X) \to L(t)^*(X) \otimes A(t)$ which determines the action of $A(t)^*$ (analogous to [10] for ordinary cohomology), and compute λ when X is such a product of projective spaces.

$L(t)^*$ is useful to formulate and prove the previous, but results can be summarized in more familiar terms.

Let $B(t) = \{f \in \mathbb{Q}[\eta] \mid \text{degree } f \leq t;\ f+(\eta^{t+1}) \in A(t)\}$,

$$B = \bigcup_{t=0}^{\infty} B(t).$$

Theorem B. (i) B is a sub-$\mathbb{Q}_p$-Hopf algebra of $\mathbb{Q}[\eta]$, where $\eta \to \eta \otimes \eta$ is the diagonal.

(ii) For each X such that $H^*(X;\mathbb{Q}_p)$ is free we have a map

$$\Lambda\colon K(X;\mathbb{Q}_p) \to K(X;\mathbb{Q}_p) \otimes B$$

which is a natural ring homomorphism and comodule map.

(iii) If $X \in K_{2n}(X;\mathbb{Q}_p)$, the $2n^{th}$ filtration subgroup, then

$$\Lambda(x) - x \otimes \eta^n \in \sum_{t=1}^{\infty} \text{Image } [K_{2n+2t}(X;\mathbb{Q}_p) \otimes \eta^n B(t)].$$

(iv) $\Lambda(x) \in K(X;\mathbb{Q}_p) \otimes [B \cap \text{Span}\{\eta^k \mid H^k(X;\mathbb{Q}_p) \neq 0\}]$.

(v) The Adams operation $\psi^k(x)$ is obtained by letting $\eta = k$ in $\Lambda(x)$.

When $p = 2$, one can show

$B(t) = \{f \in \mathbb{Q}[\eta] \mid \text{degree } f \leq t;\ 2^t f \in \mathbb{Q}_2[\eta];\ \ f(k) \in \mathbb{Q}_2$ for all units $k \in \mathbb{Q}_2^*\}$.

There are similar descriptions for $p > 2$. When $p > 2$, one can reprove the Adams splitting theorem, [1], by showing that

$$\sum_{i \in C} \pi_i \in A(t)^*, \text{ for each congruence class } C \subset \mathbf{Z}_{p-1}.$$

It is easy to check that $< \sum_{i \in C} \pi_i, f > \in \mathbb{Q}_p$ for all $f \in A(t)$.

In the following applications, take $p = 2$. If $X = S^{2n} \cup_\gamma e^{2n+2k}$, where $\gamma \in \pi_{2n+2k-1}(S^{2n})$, then $\tilde{K}(X;\mathbb{Q}_2) \cong \mathbb{Q}_2 \oplus \mathbb{Q}_2$ with generators x and y of filtrations $2n$ and $2n+2k$. By Theorem B, $\Lambda(y) = y \otimes \eta^{n+k}$, and

$$\Lambda(x) = x \otimes \eta^n + y \otimes \eta^n g,$$

where $g = a(\eta^k-1) \in B(k)$, for a rational number a which is independent of the choice of x modulo $\mathbb{Q}_2$. This gives a homotopy invariant

$$\pi_{2n+2k-1}(S^{2n}) \to (\eta^k-1)\mathbb{Q} \cap B(k)/(\eta^k-1)\mathbb{Q}_2.$$

By the previous remarks, the right side is a cyclic group of order

$$2^{\min\,[k,\nu_2(k)+2]},$$

where $\nu_2(k)$ is the largest ℓ such that 2^ℓ divides k. This invariant is essentially the 2-primary part of the complex e-invariant together with its integrality property [2].

If $n = k$ and γ had Hopf invariant one, $2^k a \sim 1$ in $\mathbb{Q}_2$ since $\psi^2(x) \equiv x^2 \not\equiv 0 \pmod 2$, so $k \leqq \nu_2(k)+2$. Hence $k = 1$, 2 or 4, reproducing essentially the proof in [3].

The Hopf invariant problem is equivalent to determining which spheres are H-spaces, and the above proof generalizes to the following result on finite dimensional H-spaces, which is not in the literature though it has been known for a few years. In particular it follows from either the methods of Hubbuck [8] or the result in [6].

<u>Theorem C</u>. For each integer $r > 0$, there exist only finitely many types $(t_1,t_2,\ldots,t_r)$ for H-spaces X of rank r such that $H^*(X)$ has no 2-torsion.

<u>Proof</u>. We have $t_i \leqq t_{i+1}$, $H^*(X;\mathbb{Q}_2)$ free, and $H^*(X;\mathbb{Q})$ an exterior algebra on generators of dimensions $2t_i-1$. Assuming the theorem false, let j be minimum such that t_j can be arbitrarily large for such X. When $t_j > 2t_{j-1}$, the jth generator is primitive. Applying the Hopf construction, we can obtain, for arbitrarily large t, spaces Y such that $\tilde{H}^k(Y;\mathbb{Q}_2)$ is free, and zero unless $2t \leq k \leqq 4t$, $H^{2t}(Y;\mathbb{Q}_2) \cong H^{4t}(Y;\mathbb{Q}_2) \cong \mathbb{Q}_2$ with generators y and y^2, and $\{k \mid H^{2k}(Y;\mathbb{Q}_2) \neq 0\}$ has at most $r+1$ elements. If $z \in K_{2t}(Y;\mathbb{Q}_2)$ corresponds to y under

$$K_{2t}(Y;\mathbb{Q}_2)/K_{2t+2}(Y;\mathbb{Q}_2) \cong H^{2t}(Y;\mathbb{Q}_2),$$

we have $z^2 \not\equiv 0 \pmod 2$. By manipulating with the diagonal in $\mathbb{Q}[\eta]$, we find that, for a given basis $\{z,z^2\} \cup \{z_\alpha \mid \alpha \in \Gamma\}$ for $\tilde{K}(Y;\mathbb{Q}_2)$,

$$\Lambda(z) = z \otimes \eta^t + \sum_{\alpha\in\Gamma} z_\alpha \otimes f_\alpha + z^2 \otimes f,$$

where (i) $f = c(\eta^{2t}-\eta^t) + \sum_{i=1}^{s} c_i(\eta^{t+k_i}-\eta^t)$; (ii) $D=\{t,t+k_1,\ldots,t+k_s,2t\}$ are the dimensions d where $H^{2d}(Y;\mathbb{Q}_2) \neq 0$; (iii) $2^t c \sim 1$ in $\mathbb{Q}_2$; (iv) $c\,\pi(\eta^d-\eta^e) \in B$ (product over all $d > e$ in D). By (iii) and (iv), 2^t divides $\pi(3^d-3^e)$, which is impossible for large t, since D has at most $r+1$ elements.

A good problem is whether one can eliminate the requirement that $H^*(X)$ have no 2-torsion in Theorem C. R. Kane [9] has made some progress in this direction. Using the methods of [4], one ought to be able to construct a coaction for <u>all</u> X,

$$\Lambda : K(X) \to K(X) \otimes C,$$

where B is possibly a proper quotient of C. Certainly $C/\text{torsion} \cong B$. This would give better proofs of Theorems A and B than the ones not given here.

Another problem is to determine enough properties of K-theory to "prove" all Adem relations for torsion free spaces. Hubbuck has proved some, [8], and one can include his results in a universal construction, starting from B, which produces a graded $\mathbb{Z}_2$-Hopf algebra D and a coaction

$$H^*(X;\mathbb{Z}_2) \to H^*(X;\mathbb{Z}_2) \otimes D.$$

D is however larger than the dual of the Steenrod algebra modulo the Bockstein.

Finally, the results of [7] on compositions of elements of non-trivial e-invariant can be improved using Theorem B. It would be useful to construct higher order K-theory operations which detect these elements directly.

References

[1] J.F. Adams, "Generalized Cohomology", Lecture Notes in Mathematics, vol. 99, Springer-Verlag, Berlin.

[2] J.F. Adams, "On the groups J(X) - IV", Topology 5 (1966), 21-71.

[3] J.F. Adams and M.F. Atiyah, "K-theory and the Hopf invariant", Quart. J. Math. (Oxford) 17 (1966), 31-38.

[4] M.F. Atiyah, "Power operations in K-theory", ibid. 165-93.

[5] P. Hoffman, "Chern character integrality for torsion free spaces", (to appear).

[6] P. Hoffman, "On the realizability of Steenrod modules", Quart. J. Math. (Oxford) (2), 20 (1969), 403-7.

[7] P. Hoffman, "Adams operations and homotopy composition", Quart. J. Math. (Oxford) (2), 19 (1968), 351-61.

[8] R. Hubbuck, "Generalized cohomology operations and H-spaces of low rank", Trans.A.M.S. 141 (1969), 335-60.

[9] R. Kane, Ph.D. Thesis, University of Waterloo, 1973.

[10] J. Milnor, "The Steenrod algebra and its dual", Ann. Math. 67 (1958), 150-71.

THE EQUIVARIANT BORDISM RING OF Z/p MANIFOLDS WITH ISOLATED FIXED POINTS

Czes Kosniowski*

S.U.N.Y., Stony Brook

1. Introduction

Let Z/p be the cyclic group of order p (p an odd prime). Let M, M' be compact closed oriented manifolds upon which the group Z/p acts with isolated fixed points (i.e. a finite-possibly zero- number of fixed points). We can form their disjoint union $M + M'$, their product $M \times M'$ and their opposites $-M$, $-M'$ (opposite orientation). In each case we get a Z/p manifold with isolated fixed points.

We say that M and M' (both of dimension m) are equivariantly bordant if there exists an oriented manifold of dimension $m + 1$ with the following two properties

(i) The boundary of W is $M - M'$.

(ii) W has a Z/p action which induces the given Z/p actions on M and M'.

This is an equivalence relation. The set of equivalence classes of Z/p manifolds with isolated fixed points forms the Equivariant Bordism Ring of Z/p manifolds with isolated fixed points. We describe this ring in section 2 for the primes 3 and 5.

---

*The author was supported in part by an N.S.F. grant and the State University of New York at Stony Brook.

By removing small equivariant neighbourhoods of each point we get a fixed point free Z/p manifold whose boundary is a collection of spheres. Corollary 3.8 states precisely which combinations of spheres with free linear Z/p actions are (fixed point free) equivariantly bordant to zero.

We only give results for $p = 3$ and 5, the general case requires more analysis and will be fully treated in [6].

Throughout, all manifolds are oriented.

2. Examples and Results

First let us describe some manifolds having $Z/3$ actions with isolated fixed points.

Example 2.1. If $\xi = \exp(2\pi i/3)$ and R the Torus $C/\{1,\xi\}$ then $z \to \xi z$ induces a $Z/3$ action on R with three fixed points.

Example 2.2. Let S be the Complex Projective Plane CP^2, with $Z/3$ action induced by

$$[z_0;z_1;z_2] \to [z_0;\xi z_1;\xi^2 z_2].$$

Theorem 2.3. If M is an m-dimensional $Z/3$ manifold with isolated fixed points, then M is equivariantly bordant to

$$a\ R^{[(m+2)/4] - [m/4]}\ S^{[m/4]} + 3N$$

where $a \in Z$ and $a = 0$ if m is odd. The symbol $[d]$ means the greatest integer less than or equal to d. $3N$ is three copies of some m-dimensional manifold N with a $Z/3$ action given by permuting the three copies of N.

We now go on to the case of $Z/5$, first some examples.

Example 2.4. Let W_1 be the Riemann Surface of genus 2 associated to the complex function $u = (z^5 - 1)^{\frac{1}{2}}$. We give it the $Z/5$ action induced by $z \to \omega^2 z$, where $\omega = \exp(2\pi i/5)$. W_1 has three fixed points -- two corresponding to $z = 0$ and one corresponding to $z = \infty$.

Let W_2 be the same surface but with $Z/5$ action induced by $z \to \omega^4 z$. (W_1 and W_2 are not equivariantly bordant.)

Example 2.5. Let X_1 be CP^2 with $Z/5$ action induced by

$$[z_0;z_1;z_2] \to [z_0;\omega z_1;\omega^2 z_2]$$

where $\omega = \exp(2\pi i/5)$.

Let X_2 also be CP^2 but with a $Z/5$ action induced by

$$[z_0;z_1;z_2] \to [z_0;\omega^2 z_1;\omega^4 z_2] .$$

Both X_1 and X_2 have three fixed points, $[1;0;0]$, $[0;1;0]$ and $[0;0;1]$.

Example 2.6. We construct a $Z/5$ manifold Y as follows

$$Y = (D^2 \times X_1) \cup_f -(D^2 \times X_2).$$

The $Z/5$ manifolds X_1 and X_2 are from example 2.5 and D^2 is the unit disc in the complex plane C with $Z/5$ action given by $z \to \omega z$. The map f is an equivariant diffeomorphism on the boundaries

$$f : \partial(D^2 \times X_1) = S^1 \times X_1 \longrightarrow \partial(D^2 \times X_2) = S^1 \times X_2$$
$$(t,[z_0;z_1;z_2]) \longrightarrow (t,[z_0;tz_1;t^2 z_2]).$$

Y is a $Z/5$ manifold with six fixed points.

Example 2.7. The Complex Projective Space CP^4 provides our last example. Let $Z = CP^4$ with $Z/5$ action given by

$$[z_0;z_1;z_2;z_3;z_4] \to [z_0;\omega z_1;\omega^2 z_2;\omega^3 z_3;\omega^4 z_4].$$

There are five fixed points.

Main Theorem 2.8. If M is an m-dimensional $Z/5$ manifold with isolated fixed points then M is equivariantly bordant to

$$\sum \alpha_{a,b,c,d,e,f}\, W_1{}^a W_2{}^b\, X_1{}^c X_2{}^d\, Y^e Z^f \;+\; 5N$$

where the sum is taken over positive integers a,b,c,d,e,f satisfying $2a + 2b + 4c + 4d + 6e + 8f = m$ and the $\alpha_{a,b,\ldots,f}$ are integers. $5N$ is five copies of some m-dimensional manifold N, with $Z/5$ action given by permuting the five copies of N.

Note that the expression in Theorem 2.8 is not uniquely defined. To get a unique expression for the case $m = 2n$, we define manifolds R_1,R_2,S_1,S_2,T and U in section 3. Then M is uniquely expressible as a sum

$$\sum_{a=0}^{(n-1)/2} \alpha_a S_1{}^a U^{[(n-2a)/4]} Q$$

$$+ \sum_{a=(n-1)/2+1}^{n} \alpha_a R_1{}^{2a-n} S_1{}^{n-a} \quad + \quad 5N$$

where $Q = R_2, S_2$ or T depending on whether $n - 2a - 4[(n-2a)/4]$ is 1, 2 or 3 respectively.

3. Proof of Theorem

We shall only prove the Main Theorem 2.8, leaving the much easier proof of Theorem 2.3 to the reader.

The normal bundle N_P to a fixed point P of the $Z/5$ manifold M is a real $Z/5$ module. The irreducible real representations of $Z/5$ are of two types

(i) one dimensional $g \to +1$

(ii) two dimensional $g \to \begin{pmatrix} \cos(2\pi ij/5) & -\sin(2\pi ij/5) \\ \sin(2\pi ij/5) & \sin(2\pi ij/5) \end{pmatrix}$

$(1 \leq j \leq 4)$

where g is the generator of $Z/5$. The one dimensional representation (i) does not occur in N_P, because N_P is normal to the fixed point P. In (ii), the representations given by $+j$ and $-j$ are equivalent, so we may restrict to the case $1 \leq j \leq 2$. Such a two dimensional real $Z/5$ module has then a canonical complex structure in which g acts as the complex number $\exp(2\pi ij/5)$. Thus N_P can be canonically written as a direct sum

$$N_P(1) \oplus N_P(2)$$

where $N_P(j)$ has a natural complex structure in which g acts as $\exp(2\pi ij/5)$. (In particular if M has fixed points then M must be of even dimension.) This complex structure on N_P induces an orientation on N_P and hence on P -- this may or may not agree with the orientation on P induced from $TM|_P$. We record this fact by saying $or(P) = 1$ if the orientations agree and $or(P) = -1$ otherwise.

This decomposition of the normal bundle can be used to define a map γ from the equivariant bordism ring of $Z/5$ manifolds with isolated fixed points to the polynomial ring $Z[x_1, x_2]$ in two variables. If $d(j,P)$ is the complex dimension of $N_P(j)$ then

$$\gamma\colon M \to \sum \text{ or}(P)\ x_1^{d(1,P)}\ x_2^{d(2,P)}$$

where the sum is taken over the fixed points $\{P\}$ in M.

It is trivial to check that γ is a ring homomorphism whose kernel is precisely the equivariant bordism classes of $Z/5$ manifolds with no fixed points. This set is exactly $5\Omega_*$, where Ω_* is the oriented bordism ring. (This follows from [4;36.5], where Conner and Floyd determine the ring $\Omega_*(BZ/5)$. This ring is the bordism ring of free $Z/5$ manifolds in which bordisms go through free $Z/5$ manifolds. If we allow bordisms with fixed points then it is easy to see that we get $5\Omega_*$.) If $5N \in 5\Omega_*$, then $5N$ is five copies of some manifold N with a $Z/5$ action given by permuting the five copies of N.

If we denote by Ω^5_* the equivariant bordism ring of $Z/5$ manifolds with isolated fixed points, then $\Omega^5_*/5\Omega_*$ is isomorphic, via γ, to a subring of $Z[x_1,x_2]$. Call this subring B, it is a graded ring $B = \sum B_N$ $(n \geq 1)$, where if M is $2n$ dimensional then $\gamma(M)$ lies in B_n.

The examples in section 2 have the following image under γ:

Lemma 3.1. $\gamma(W_1) = -x_1 + 2x_2$; $\gamma(W_2) = -2x_1 - x_2$

$\gamma(X_1) = -x_1^2 + 2x_1x_2$; $\gamma(X_2) = -2x_1x_2 - x_2^2$

$\gamma(Y) = x_1^3 + 4x_1^2x_2 + x_1x_2^2$; $\gamma(Z) = 5x_1^2x_2^2$.

The proof is simple.

The manifolds W_1, W_2, X_1, X_2, Y and Z generate a subring of $\Omega^5_*/5\Omega_*$, denote its image in $Z[x_1,x_2]$ by $C = \sum C_n$.

Clearly $C \subset B$.

If we denote $-x_1 + 2x_2$ by x and x_2 by y then $Z[x_1,x_2]$ is isomorphic to $Z[x,y]$. Consider the following $Z/5$ manifolds with isolated fixed points:

$R_1 = W_1$	$\gamma(R_1) = x$
$R_2 = -2W_1 - W_2$	$\gamma(R_2) = 5y$
$S_1 = -W_1W_2 - 2X_1$	$\gamma(S_1) = xy$
$S_2 = 2S_1 - X_2$	$\gamma(S_2) = 5y^2$

$T = -W_2S_2 - 2W_1X_1 - 2Y$ $\qquad\qquad \gamma(T) = 5y^3$

$U = S_2(W_1{}^2 - X_2 - 2S_1) - Z$ $\qquad\qquad \gamma(U) = 5y^4$.

The manifolds R_1, R_2, S_1, S_2, T and U generate a subring of $\Omega_*^5/5\Omega_*$, denote its image in $Z[x_1,x_2]$ by $D = \sum D_n$.

Clearly $D \subset C \subset B$. We shall show eventually that all three rings are isomorphic.

Lemma 3.2.

$$D_n = \{\textstyle\sum_{a=0}^{n} \alpha_a 5^{r(a,n)} x^a y^{n-a} ; \alpha_a \in Z\}$$

where $r(a,n)$ is $[(n-2a-1)/4] + 1$ if $a \le (n-1)/2$ and is zero otherwise.

Proof. If $2a \ge n$ then $x^a y^{n-a} = x^{2a-n}(xy)^{n-a}$ lies in D_n. If $2a < n$ then $x^a y^{n-a} = (xy)^a y^{n-2a}$, and so $5^{r(a,n)} x^a y^{n-a}$ lies in D_n.

Now, D_n is a $\mathbb{Z}$ module of rank $n + 1$. In general if E is a Z module of rank $n + 1$, we may think of E as a linear transformation of Z^{n+1}. We denote the determinant of this transformation by det E.

Lemma 3.3.

$$\det D_n = \Pi_{a=0}^{[(n-1)/2]} 5^{r(a,n)}$$

This follows from Lemma 3.2.

Let us return to the ring $B = \sum B_n$.

Theorem 3.4. If $\sum_{a=0}^{n} \alpha_a x_1{}^a x_2{}^{n-a}$ lies in B_n then

$$\sum \alpha_a f(\underbrace{(\omega-1)^2,..,(\omega-1)^2}_{a}, \underbrace{(\omega^2-1)^2,..,(\omega^2-1)^2}_{n-a}) \{(\omega+1)/(\omega-1)\}^a \{(\omega^2+1)/(\omega^2-1)\}^{n-a}$$

lies in $\mathbb{Z}[\omega]$, where $\omega = \exp(2\pi i/5)$ and f is any symmetric homogeneous polynomial in n variables, (of degree $k < n$).

This theorem follows quite easily from various works -- see [5;§2], [3;§8], [1;§5], [2].

For applications, Theorem 3.4 is inconvenient, we express it in a more convenient form in the next lemma.

Lemma 3.5. If $5^r \sum \beta_a x_1{}^a x_2{}^{n-a}$ lies in B_n then

$$\sum\beta_a \ f(1,1,..,1,4,4,..,4) \ / \ 2^{n-a} = 0 \bmod 5$$

for all symmetric homogeneous polynomials f of degree k, with $k \leq [(n-1-4r)/2]$.

Proof. Multiply the equation in Theorem 3.4 by $(\omega-1)^{n-2k-4r}$, where k is the degree of f. Note that $5^r = (\omega-1)^{4r} \times$ unit. The result follows by reducing modulo $(\omega-1)$, recalling that $(\omega^j-1)/(\omega-1) \bmod (\omega-1)$ is $j \bmod 5$.

For our purposes it will be sufficient to consider the q-th elementary symmetric functions δ_q. Define

$$\delta(q,a) = \delta_q(\underbrace{1,1,..,1}_{a},\underbrace{4,4,..,4}_{n-a}) \ / \ 2^{n-a} .$$

So, if $5^r\sum \beta_a \ x_1{}^a \ x_2{}^{n-a}$ lies in B_n then

$$\sum \beta_a \ \delta(q,a) = 0 \bmod 5 \quad \text{for} \quad q = 0,1,......,[(n-1-4r)/2].$$

Lemma 3.6. There is a basis $\{e_a \ ; \ a = 0,1,......,n\}$ of Z^{n+1} such that B_n is contained in the submodule with basis $\{5^{r(a,n)}e_a \ ; \ a = 0,1,......,n\}$.

Proof. First we observe that the vectors $\sum \delta(q,a) \ x_1{}^a \ x_2{}^{n-a} \in Z^{n+1} \otimes Z/5$ $(q = 0,1,...., (n-1)/2)$ are linearly independent over $Z/5$.

Find a basis in $Z^{n+1} \otimes Z/5$ for the sub-vector space orthogonal to the vectors $\sum \delta(q,a) \ x_1{}^a \ x_2{}^{n-a}$. (Two vectors $\sum \alpha_a \ x_1{}^a \ x_2{}^{n-a}$ and $\sum \beta_a \ x_1{}^a \ x_2{}^{n-a}$ are orthogonal if $\sum \alpha_a\beta_a = 0$.) Let this basis be

$$\{d_{[(n-1)/2]+1},...........,d_n\}.$$

We may consider these vectors as elements in Z^{n+1}, call these d_i' . By taking multiples of the d_i, $i = [(n-1)/2] + 1,....,n$ we get elements d_i' which can be extended to a basis of Z^{n+1} -- say

$$\{c_i \ ; \ i = 0,1,.....,[(n-1)/2]\}$$

$$\{d_i' \ ; \ i = [(n-1)/2] + 1,......,n\}$$

Then B_n is contained in the submodule with basis

$$\{d_i' \ ; \ [(n-1)/2] < i \leq n\} \ \text{and} \ \{5c_i \ ; \ 0 \leq 1 \leq [(n-1)/2]\}.$$

We let $e_i = d_i'$ ($[(n-1)/2] < i \leq n$), then repeat the argument on the submodule with basis $\{c_i\}$ using the vectors

$\sum \delta(q,a)\ x_1^{\ a}\ x_2^{\ n-a}$ for $q = 0,1,\ldots\ldots,[(n-1-4)/2]$. We continue in this manner until we get the desired result.

Let A_n be the submodule of Z^{n+1} with basis $\{5^{r(a,n)} e_a\}$.

We have now $D_n \subset C_n \subset B_n \subset A_n$, and from the definition of A_n, we also have $\det A_n = \det D_n$. This proves that $A_n = B_n = C_n = D_n$ and hence proves the Main Theorem 2.8.

Theorem 3.7.

$A_n = B_n = C_n = D_n = \{ \sum_{a=0}^{n} \alpha_a 5^{r(a,n)} (-x_1+2x_2)^a\ x_2^{\ n-a} ; \alpha_a \in Z\}$.

If an oriented sphere has a free linear $Z/5$ action, then we can extend it to an action of R^{2n} (where $2n-1$ is the dimension of the sphere), so we get a real $Z/5$ module and hence a monomial

$$\gamma(S) = \mathrm{or}(S)\ x_1^{\ d(1,S)}\ x_2^{\ d(2,S)} .$$

Corollary 3.8. If $S_1, S_2, \ldots\ldots, S_k$ is a collection of oriented spheres of dimension $2n-1$, each with a free linear $Z/5$ action then the disjoint union $\sum S_i$ is equivariantly bordant (through fixed point free $Z/5$ manifolds) to zero (i.e. $\sum S_i = 0$ in $\Omega_*(BZ/5)$) if and only if

$$\sum_{i=1}^{k} \gamma(S_i) = \sum_{a=0}^{n} \alpha_a\ 5^{r(a,n)}\ (-x_1+2x_2)^a\ x_2^{\ n-a}$$

for some $\alpha_a \in Z$.

We leave the proof to the reader.

References

[1] M.F. Atiyah and G.B. Segal, Equivariant K-theory, University of Warwick Notes.

[2] M.F. Atiyah and G.B. Segal, The index of elliptic operators: II, Ann. of Math. 87(1968), 531-545.

[3] M.F. Atiyah and I.M. Singer, The index of elliptic operators: III, Ann. of Math. 87(1968), 546-604.

[4] P.E. Conner and E.E. Floyd, Differentiable Periodic Maps, Springer, 1964.

[5] C. Kosniowski, What the fixed points say about a Z/p manifold, To appear in Journal of the London Math. Society.

[6] C. Kosniowski, Z/p manifolds with isolated fixed points, To appear.

THE KERNEL OF THE LOOP MAP

David Kraines

Duke University, Durham

Let X be a 1-connected H-space with $H^*(X;Z_p)$ of finite type. The loop map is the loop suspension homomorphism

$$\sigma\colon QH^q(X;Z_p) \to PH^{q-1}(\Omega X;Z_p).$$

Let Ψ_r be the secondary cohomology operation based on the Adem relation

$$\mathcal{P}^{np^r-1}\,\mathcal{P}^{np^{r-1}}\cdots\mathcal{P}^n = 0$$

and let β_k be the p^k-th order Bockstein.

<u>Theorem</u>. Assume that $x \in \operatorname{Ker}\sigma$. Either there is an indecomposable class $u \in H^{2m+1}(X;Z_p)$ such that

$$\beta_k\,\mathcal{P}^{mp^{k-1}}\cdots\mathcal{P}^m u$$

is defined and contains x or else there is an indecomposable class $v \in H^{2n}(X;Z_p)$ such that

$$\beta_k\,\mathcal{P}^{(np^r-1)p^{k-1}}\cdots\mathcal{P}^{(np^r-1)p}\,\Psi_r(v)$$

is defined and contains x.

MOTION OF A PARTICLE IN A TOPOLOGICAL SPACE

R.G. Lintz

McMaster University, Hamilton

Some years ago the idea of derivative in a topological space had been introduced by the author and with further development of this theory in collaboration with other people we arrived at the point of starting the study of motion in topological spaces. Details can be seen in: "The concept of differential equation in topological spaces and generalized mechanics", by R.G. Lintz and V. Buonomano, to appear soon in J. für die Reine und Angew. Math.

One starts by defining a Gauss structure in a topological space X and particles in X are identified with particular kinds of g-functions. Four principles are postulated to govern the motion of particles in X and several conclusions are drawn from them. A strong connection with Schrödinger's wave functions is discussed and so far two simple cases of differential equations have been studied which are relevant to these questions. All these concepts are too involved to be explained in a few words, but they are discussed in detail in the paper referred to above.

$\text{Ext}_{Ap}^{2,*}$ (Zp,Zp) AND RELATIONS IN THE STEENROD ALGEBRA

Stavros Papastavridis

Brandeis University

Chapter 1. Introduction

It is well known that computing $\text{Ext}_{Ap}^{2,*}$ (Zp,Zp) means in a sense, find a minimal set of relations among the generators of the Steenrod algebra Ap, the connection was formalized by C.T.C. Wall, see [9]. The computation of $\text{Ext}_{Ap}^{2,*}$ (Zp,Zp) has been done a long time ago by J.F. Adams [1], when $p = 2$, and independently by A. Liulevicius [3], N. Shimada [5] and T. Yamanoshita [6], when p is an odd prime, and those computations were key steps in settling the Hopf-invariant conjecture. I have the feeling that something is missing from the whole picture, specifically those proofs don't provide a somehow explicit minimal set of relations when p is an odd prime, and also they don't give a practical way that enables you, given a specific relation in Ap, to analyze it to a linear combination of relations from a given minimal set. When $p = 2$ C.T.C. Wall, [9], computed $\text{Ext}_{Ap}^{2,*}$ (Zp,Zp) in a more direct way, he provided a more or less specific minimal set of relations for the generators of A_2, and his proof gives a practical algorithm that enables you to analyze a given relation in A_2 to a linear combination of simpler relations from the specified minimal set. The importance of that lies in the fact that relations in Ap correspond to secondary cohomology operations and so this provides a way to analyze a secondary cohomology operation to simpler ones which hopefully it's easier to compute. Such a case was my work on the Arf-invariant [4], where we had the case of a secondary cohomology operation detecting the Arf-invariant, under certain conditions this could be split in simpler secondary cohomology operations which were trivially zero. That motivated me to look for something analogous for the mod-p case where p is an odd prime. From now on in the present work p is going to be an odd prime, Ap is the mod-p Steenrod algebra and A'p is the algebra of reduced power operations which of course is a subalgebra of Ap. Our analysis applies equally well in the mod-2 case with minor modifications in the notation.

Definition 1.1. Let A be the free associative graded algebra over Zp with unit generated by the symbols $\{P^i\}$ $i \geq 1$, and δ, where grade $P^i = 2(p-1)i$ grade $\delta = 1$, let A' be the subalgebra of A generated by $\{P^i\}$ $i \geq 1$, let B be the subalgebra of A generated by $\{P^{p^i}\}$ $i \geq 0$, and δ, and finally let B' be the subalgebra of B generated by $\{P^{p^i}\}$ $i \geq 0$. Let f be the obvious homomorphism of graded algebras over Zp: A $\to$ Ap which associates to P^i the i-th reduced power operation, and to δ the well-known Bockstein coboundary operator associated with the exact coefficient sequence

$$0 \to Zp \to Zp^2 \to Zp \to 0$$

It is well-known (by definition), see [6], that the kernel of f is an ideal over A generated by the elements: $R(a,b) = P^aP^b - \sum_{t=0}^{[a|p]}(-1)^{a+t}\binom{(p-1)(b-t)-1}{a-pt} P^{a+b-t}P^t$ when $0 < a < pb$,

$$R'(a,b) = P^a\delta P^b - \sum_{t=0}^{[a|p]}(-1)^{a+t}\binom{(p-1)(b-t)}{a-pt}\delta P^{a+b-t}P^t$$

$$- \sum_{t=0}^{[(a-1)|p]}(-1)^{a+t-1}\binom{(p-1)(b-t)-1}{a-pt-1}P^{a+b-t}\delta P^t$$

when $0 < a \leq pb$.

The main results of this paper are the following two theorems.

Theorem 1.2. The following elements of A are a minimal set of ideal generators for the kernel of f :

a) $R(p^i, (p-1)p^i)$ $1 \geq 0$

b) $R(p^{i+1}, 2p^i)$ $i \geq 0$

c) $R(p^{i+1}, (p+1)p^i)$ $i \geq 0$

d) $R(p^{i+1}, (1+p^S-p)p^i$ $i \geq 0$, $S \geq 2$

e) $R'(1,1)$

f) $R'(1, p^i-1)$ $i \geq 1$

g) δ^2

h) $R(p^i, (mp+K)p^i)$ $1 \geq 0$, $m \geq 0$, $0 \leq K \leq p - 2$.

Theorem 1.3. (Liulevicius - Shimada - Yamanoshita). $\mathrm{Ext}^{2,*}_{Ap}(Zp,Zp)$ is a graded vector space over Zp, and it has a basis with exactly one element in each one of the following dimensions:

a) $2(p-1)p^{i+1}$ $i \geq 0$

b) $2(p-1)(p^{i+1} + 2p^i)$ $i \geq 0$

c) $2(p-1)(2p^{i+1} + p^i)$ $i \geq 0$

d) $2(p-1)(p^i + p^{S+i})$ $i \geq 0$ $S \geq 2$

e) $4(p-1) + 1$

f) $2(p-1)p^i + 1$ $i \geq 1$

g) 2

The paper is arranged as follows. In Chapter 2, we prove that the elements described in Theorem 1.2 form a minimal set of ideal generators, and in Chapter 3 we prove that they generate the kernel of f.

Chapter 2.

Definition 2.1. Let us call the set of elements of A described in Theorem 1.2 part a) X_a, those of part b) X_b, ... etc. those of part h) X_h. Define $X = \bigcup X_i$ where $i = a,b, \ldots h$, and $Y = X - X_h$. Obviously $X_i \cap X_j = \emptyset$ $i \neq j$, $i,j = a,b, \ldots, h$. We call monomial in A a product of the elements $\{P^i\}$ $i \geq 1$ and δ. We call polynomial in A a linear combination of monomials with coefficients in Zp. We call degree of decomposability of a monomial in A the number of its factors. We call degree of decomposability of a polynomial in A the lowest degree of decomposability of its monomials. We call an element of A decomposable if it has a polynomial expression with degree of decomposability greater or equal than two.

Next we state a few well-known results.

Lemma 2.2. Let $a = \sum_{i=0}^{m} a_i p^i$ and $b = \sum_{i=0}^{m} b_i p^i$, $0 \leq a_i, b_i < p$, then $\binom{a}{b} = \Pi_{i=0}^{m} \binom{a_i}{b_i} \bmod p$.

Proof. See [6] page 5.

Lemma 2.3. If $0 \leq k \leq p-2$, and $m \geq 0$ then $R(p^i, (mp+k)p^i) = -(k+1)P^{mp^{i+1}+(k+1)p^i} + P^{p^i} P^{(mp+k)p^i} - \sum_{t=1}^{p^{i-1}} (-1)^{p^i+t}$

$$\binom{(p-1)(mp^{i+1}+kp^i-t)-1}{p^i-pt} \circ P^{mp^{i+1}+(k+1)p^i-t} P^t$$

Proof. By the definition of $R(p^i, (mp+k)p^i)$ (see Introduction) all we have to prove is that

$$\binom{(p-1)(mp^{i+1}+kp^i)-1}{p^i} = -(k+1) \bmod p .$$

Really we distinguish two cases whether $k = 0$ or $\neq 0$, and by applying the previous lemma we get:

Case 1. $k = 0$, then $m > 0$ otherwise the symbol $R(p^i, (mp+k)p^i)$ becomes meaningless, and $(p-1)(mp^{i+1}+kp^i)-1 = [(p-1)m-1]p^{i+1} + p^{i+1} - 1$

$= [(p-1)m-1]p^{i+1} + (p-1)p^i + (p-1)p^{i-1} + \ldots + (p-1)p + (p-1)$

so $$\binom{(p-1)(mp^{i+1}+kp^i)-1}{p^i} = \binom{p-1}{1} = p-1 = -1 \quad \bmod p.$$

Case 2. $k > 0$, then

$(p-1)(mp^{i+1}+kp^i)-1 = (p-1)mp^{i+1} + [(p-1)k-1]p^i + p^i - 1 = (p-1)mp^{i+1}$

$+ [(p-1)k-1]p^i + (p-1)(p^{i-1} + \ldots + \ldots + p+1) =$

and so by the previous lemma

$$\binom{(p-1)(mp^{i+1}+kp^i)-1}{p^i} = \binom{(p-1)k-1}{1} = -(k+1) \quad \bmod p.$$

Remark. Observe that under the conditions of the previous lemma $(k+1) \nmid 0 \bmod p$ so

$$P^{mp^{i+1}+(k+1)p^i} = -\frac{1}{k+1} R(p^i, (mp+k)p^i) + \text{decomposables}.$$

Proposition 2.4. Suppose that a has p-adic expansion $\sum a_i p^i$, $0 < a_i < p$, and let I be the set of indexes i which appear in the expansion, we put on I the natural order of integers, then $P^a = \sum r\,Rq + \Pi_{i \in I} (1/(a_i)!)(P^{p^i})^{a_i} + \ldots +$ (element of B' with degree of decomposability strictly greater than $\sum_{i \in I} a_i$), r,q are elements of B' and R are non-empty products of elements of X_h, (see Definition 2.1).

Proof. We proceed by induction on a. Suppose that the statement of the proposition is true for all natural numbers which are smaller than a.

If a is a power of p the proposition is obvious, if a is not a power of p then $a = \sum_{i=k}^{m} a_i p^i$, with $0 \leq a_i < p$, $a_k > 0$, then by applying the previous lemma

$$P^a = -(1/a_k)R(p^k, a-p^k) + (1/a_k)P^{p^k} P^{a-p^k} - (1/a_k)\sum_{t=1}^{p^{k-1}} \binom{(p-1)(a-p^k-t)-1}{p^k - p^t}$$

$P^{a-t} P^t$ and the result follows by the inductive assumption.

Definition 2.5. For any element $y \in Y$ in the specific polynomial form given in the introduction (cf. also Definition 2.1), by making substitutions using the formula of the previous Proposition and by executing the obvious reductions, one gets an expression

$$y = R_y + \sum_r a_r\, r\, b_r, \quad \text{where } a_r, b_r, R_y \in B$$

and r ranges over all non-empty products of elements of X_h.

Proposition 2.6. a) If $y = R(p^i, (p-1)p^i)$, $i \geq 0$, then $R_y = (1/(p-1)!)(P^{p^i})^p$ + (elements of B' of higher decomposibility);

b) if $y = R(p^{i+1}, 2p^i)$, $i \geq 0$, then $R_y = (1/2)P^{p^{i+1}} P^{p^i} P^{p^i} + (1/2)P^{p^i} P^{p^i} P^{p^{i+1}} - P^{p^i} P^{p^{i+1}} P^{p^i}$ + (elements of B' of higher decomposability);

c) if $y = R(p^{i+1}, (p+1)p^i)$, $i \geq 0$, then $R_{y_i} = P^{p^{i+1}} P^{p^i} P^{p^{i+1}} - (1/2)P^{p^i} P^{p^{i+1}} P^{p^{i+1}} - (1/2)P^{p^{i+1}} P^{p^{i+1}} P^{p^i}$ + (elements of B' of higher decomposability);

d) if $y = R(p^{i+1}, (1+p^S-p)p^i)$, $i \geq 0$, $S \geq 2$, then $R_y = P^{p^i} P^{p^{i+S}} - P^{p^{i+S}} P^{p^i}$ + (elements of B' of higher decomposibility);

e) if $y = R'(1,1)$, then $R_y = y$;

f) if $y = R'(1, p^i - 1)$, then $R_y = \delta P^i - P^i \delta$ + (elements of B of higher decomposability), $i \geq 1$;

g) if $y = \delta^2$ then $R_y = y = \delta^2$.

Proof. This follows easily by applying Proposition 2.4 and looking carefully at the definitions of the R's.

Proposition 2.7. It is not possible to have an equality of the form $R_{y_0} = \sum_{y \neq y_0} r_y R_y q_y$, $y_0 y \in Y$, $r_y, q_y \in B$.

Proof. We distinguish whether $y_0 \in X_a$, or X_b, ..., or X_g. The idea is that each R_{y_0} contains characteristic elements, specifically the elements of least decomposability (see previous proposition), which cannot be reproduced by a combination of R_y's, $y \neq y_0$. We leave the details as an exercise. Essentially this Proposition means that the R_y's are "ideally" independent over B, subsequently we are going to prove that they generate the kernel of $f|B$.

Lemma 2.8. The elements of X_h and $\{P^{p^i}\}$, $i \geq 0$ and δ, are free generators for the algebra A.

Proof. Follows easily from Lemma 2.3.

Proposition 2.9. It is not possible to have an equality of the form

$$x_0 = \sum_{x \neq x_0} r_x x \; q_x \quad ; \quad x, x_0 \in X, \quad r_x, q_x \in A.$$

Proof. Suppose that we have such an equality. It is impossible to have $x_0 \in X_h$, because in this case x_0 contains an indecomposable which cannot appear on the other side of the equality. So let us assume that $x_0 \in Y$. Making substitution using Proposition 2.4 we get

$$R_{x_0} + \sum_r a_r \, r \, b_r = \sum_{y \in Y} c_y R_y d_y + \sum_r g_r \, r \, f_r \ ,$$

where $a_r, b_r, c_y, d_y, g_r, f_r \in B$ and r ranges over all non-empty products of elements of X_h. But by the previous Lemma this means $R_{x_0} = \sum_{\substack{y \in Y \\ y \neq x_0}} r R_y q$ which is impossible by the previous Proposition.

What we just proved means that the elements of X are "ideally" independent over A, our main task is to prove that they generate the kernel of f. Last Proposition was conjectured by the late N. Steenrod in his last paper [7], p. 89. So with Proposition 2.7 and 2.9 we finished half of the task, by proving that the elements of X and $\{R_y\}$

are minimal ideal generators for the algebras A and B correspondingly; it now remains to show that they do generate the kernels of f and $f|B$ correspondingly.

Chapter 3.

Definition 3.1. If $a, b \in A$ then $[a,b] = ab - ba$. We define inductively the elements Q_i as follows $Q_o = \delta$, $Q_{i+1} = [P^{p^i}, Q_i]$. We introduce a relation among elements of A as follows. Suppose $a,b \in A$ are written as polynomial expressions in reduced form, then $a \sim b$ means $a = b + c + d$ where c belongs to the ideal generated by the elements of X and d has degree of decomposability greater than that of a and b. Obviously $\sim$ is a reflexive and symmetric relation compatible with multiplication.

Our first task will be to prove the following statement.

Proposition 3.2. Every element of A is equivalent $\sim$ with a linear combination of elements of the form $Q_o^{\varepsilon_o} \dots Q_n^{\varepsilon_n} x$, where $x \in A'$, $\varepsilon_o, \dots, \varepsilon_n = 0$ or 1.

Proof. It follows immediately from the following six Lemmas.

Lemma 3.3. If $a, b, c \in A$, then

$[a,[b,c]] + [c,[a,b]] + [b,[c,a]] = 0$,

and if $[a,c] \sim 0$ then $[a,[b,c]] \sim [[a,b],c]$.

Proof. Trivial.

The next definition will simplify considerably the notation.

Definition 3.4. Put $S_i = P^{p^i}$, $i \geq 0$

Lemma 3.5. a) $S_i^p \sim 0$

b) $[[S_{i+1},S_i],S_i] \sim 0$

c) $[S_{i+1},[S_{i+1},S_i]] \sim 0$

d) $[S_i,S_{i+s}] \sim 0$, $s \geq 2$

e) $[S_o,[S_o,\delta]] \sim 0$

f) $[S_i,\delta] \sim 0$ $i \geq 1$

g) $[\delta,[\delta,S_o]] \sim 0$.

Proof. A direct translation of Proposition 2.6 in the notation just introduced.

Lemma 3.6. $[S_j, Q_i] \sim 0$, if $i \neq j$.

Proof. We distinguish two cases whether $i > j$ or $j > i$.

Case 1. Suppose $j > i$. Then S_j, "commutes" for our purposes with δ and S_k for $k < i$ by the previous proposition, so $[S_j, Q_i] \sim 0$.

Case 2. Suppose $j < i$, we proceed by induction on the difference $i - j$. First we are going to prove separately the lemma for $i - j = 1$ and $i - j = 2$.

Suppose $j \geq 1$, then

$$\begin{aligned}
[S_j, Q_{j+1}] &= [S_j, [S_j, Q_j]] \\
&= [S_j\ [S_j, [S_{j-1}, Q_{j-1}]]\ ] \\
&\sim [S_j, [[S_j, S_{j-1}], Q_{j-1}]] \quad \text{by Case 1, and Lemma 3.3} \\
&\sim [[S_j, [S_j, S_{j-1}]], Q_{j-1}] \quad \text{same reason} \\
&\sim 0 \quad \text{by Proposition 3.5. c).}
\end{aligned}$$

Suppose $j = 0$, then

$$[S_j, Q_{j+1}] = [S_0, [S_0, \delta]\ ] \sim 0 \quad \text{by Lemma 3.5 e).}$$

Next we will prove that $[S_j, Q_{j+2}] \sim 0$, really

$$\begin{aligned}
[S_j, Q_{j+2}] &= [S_j, [S_{j+1}, [S_j, Q_j]]\ ] \\
&\sim [S_j, [\ [S_{j+1}, S_j], Q_j]] \quad \text{by the previous result and Lemma 3.3} \\
&\sim -[S_j, [Q_j, [S_{j+1}, S_j]\ ]] \\
&\sim -[[S_j, Q_j], [S_{j+1}, S_j]] \\
&\sim -[[S_j, S_{j+1}], [S_j, Q_j]] \\
&\sim -[[S_j, S_{j+1}], Q_{j+1}] \\
&\sim -[S_j, [S_{j+1}, Q_{j+1}]] \quad \text{by the previous result} \\
&\sim -[S_j, Q_{j+2}], \quad \text{so}
\end{aligned}$$

$2[S_j, Q_{j+2}] \sim 0$, so $[S_j, Q_{j+2}] \sim 0$.

Next, suppose that $[S_j, Q_{j+k}] \sim 0$ for some $k \geq 2$, then

$$\begin{aligned}
[S_j, Q_{j+k+1}] &= [S_j, [S_{j+k}, Q_{j+k}]\ ] \\
&\sim [\ [S_j, S_{j+k}], Q_{j+k}], \quad \text{by the inductive assumption,} \\
&\sim 0, \quad \text{by Lemma 3.5 d), and this ends the proof.}
\end{aligned}$$

Lemma 3.7. $[Q_i,\delta] \sim 0$.

Proof. For $i = 0,1$ it's Lemma 3.5. Suppose now that the lemma is true for some value $i \geq 1$ then $[Q_{i+1},\delta] = [\,[S_i,Q_i]\delta]$
$\sim [S_i,[Q_i,\delta]\,]$, by Lemma 3.5. f),

~ 0, by inductive assumption.

Lemma 3.8. $[Q_i,Q_j] \sim 0$.

Proof. For $i = j$ it's obvious, we can assume without losing generality that $i > j$, we keep i fixed and we apply induction on j. In first place $[Q_i,Q_0] \sim 0$ by the previous Lemma.

Assume $[Q_i,Q_j] \sim 0$ and that $j + 1 < i$ then
$[Q_i,Q_{j+1}] = [Q_i,[S_j,Q_j]\,]$
$\sim [\,[Q_i,S_j],Q_j]$, by the inductive assumption,
~ 0, by Lemma 3.6.

Lemma 3.9. $Q_i\ Q_i \sim 0$.

Proof. We proceed by induction on i. If $i = 0$ this is obvious. Suppose that $Q_i\ Q_i \sim 0$; then

$$\begin{aligned}
Q_{i+1}Q_{i+1} &= [S_i,Q_i]\ [S_i,Q_i] \\
&= [S_iQ_i - Q_iS_i]\ [S_iQ_i - Q_iS_i] \\
&= (S_iQ_i)^2 + Q_iS_iQ_iS_i - Q_iS_iS_iQ_i - S_iQ_iQ_iS_i \\
&\sim (S_iQ_i)^2 + Q_iS_i\ [Q_i,S_i] \\
&\sim (S_iQ_i)^2 - Q_iS_iQ_{i+1} \\
&\sim (S_iQ_i)^2 - Q_iQ_{i+1}S_i \quad \text{(by Lemma 3.6.)} \\
&\sim (S_iQ_i)^2 - Q_i\ (S_iQ_i - Q_iS_i)S_i \\
&\sim (S_iQ_i)^2 - (Q_iS_i)^2 + Q_iQ_iS_iS_i \\
&\sim (S_iQ_i)^2 - (Q_iS_i)^2 \quad \text{(by inductive assumption).}
\end{aligned}$$

In an analogous way

$$\begin{aligned}
Q_{i+1}Q_{i+1} &\sim (Q_iS_i)^2 + S_iQ_iS_iQ_i - Q_iS_iS_iQ_i \\
&\sim (Q_iS_i)^2 + [S_i,Q_i]\ S_iQ_i \\
&\sim (Q_iS_i)^2 + S_i\ [S_i,Q_i]\ Q_i \\
&\sim (Q_iS_i)^2 + S_i\ (S_iQ_i - Q_iS_i)Q_i
\end{aligned}$$

$\sim (Q_iS_i)^2 - (S_iQ_i)^2$, the combination of those two calculations gives $Q_{i+1}Q_{i+1} \sim 0$.

That was the last of the six lemmas that we had promised after Proposition 3.2.

And now we are ready to prove Theorem 1.2.

Proof of Theorem 1.2. Because of Proposition 2.9 it is enough to prove that every element in the kernel of f belongs to the two-sided ideal generated by the elements of X. Suppose $\omega \varepsilon A$, because in given degree there is only a finite number of elements of A, by the previous proposition, ω must have the form

$$\omega = \sum_\varepsilon Q_0^{\varepsilon_0} \ldots Q_n^{\varepsilon_n} x_\varepsilon + \sum_{x\varepsilon X} r_x\, x\, q_x$$

where $x_\varepsilon \; \varepsilon \; A'$ and $\varepsilon = (\varepsilon_0, \ldots \varepsilon_n)$ is a n-tuple of 0's and 1's. If $f(\omega) = 0$ then $\sum_\varepsilon Q_0^{\varepsilon_0} \ldots Q_n^{\varepsilon_n} f(x_\varepsilon) = 0$ but by Milnor's Theorem, (see [2], p. 163, Th. 4a) this means that $f(x_\varepsilon) = 0$. But because of Steenrod's result, (see [7] p. 92, Th. 6.2) this means that x_ε belongs to the ideal generated by the elements of X, which establishes the theorem.

Remark. In Steenrod's result used in the last proof, there is a typographical error, specifically in the statement of the theorem he doesn't mention $R(p^{i+1},(p+1)p^i)$, but looking at the proof it is clear that the omission is typographical.

Also Proposition 3.2, of course, is well-known, see [2], but we wanted to give an independent "constructive" proof.

Theorem 3.10. The elements $R_y \; y \; \varepsilon \; Y$ form a minimal set of generators for the kernel of $f|B$.

Proof. By Definition 2.5 and Proposition 2.7 it is enough to prove that they generate $\ker (f|B)$. Suppose $\omega \; \varepsilon \; B$ and $f(\omega) = 0$ then by Theorem 1.2 $\omega = \sum r\, R_q$, where $r,q \; \varepsilon \; A$, $R \; \varepsilon \; X$; now making substitutions in the second member of the above equality using Proposition 2.4 we get

$\omega = \sum_{y\varepsilon Y} r_y\, R_y + \sum r\, R_q$, where $r,q\; r_y, q_y \; \varepsilon \; B$, and R is non-empty product of elements of X_h, but by Lemma 2.8 this means $\omega = \sum_{y\varepsilon Y} r_y\, R_y\, q_y$.

And finally we reprove Liulevicius-Shimada-Yamanoshita's Theorem.

Proof of Theorem 1.3. This is immediate by the previous Theorem, and Wall's result (see [9] p. 440).

References

[1] J.F. Adams, On the non-existence of elements of Hopf invariant one, Annals of Math. 72(1960), 20-104.

[2] J. Milnor, The Steenrod algebra and its dual, Annals of Math. 67(1958), 150-171.

[3] A. Liulevicius, The factorization of cyclic reduced powers by secondary cohomology operations, Memoirs of AMS, number 42.

[4] S. Papastavridis, The Arf invariant of manifolds with few non-zero Stiefel-Whitney classes, (to appear).

[5] N. Shimada, Triviality of the mod-p Hopf invariant, Proc. Japan. Acad. 36(1960), 68-69.

[6] N. Steenrod & D.B.A. Epstein, Cohomology operations, Annals of Math. Studies 50.

[7] N. Steenrod, Polynomial algebras over the algebra of cohomology operations, in Lecture Notes in Mathematics, Springer-Verlag, 1972, number 196.

[8] T. Yamanoshita, On the mod-p Hopf invariant, Proc. Japan. Acad. 36(1960), 97-98.

[9] C.T.C. Wall, Generators and relations in the Steenrod algebra, Annals of Math. 72(1960), 429-444.

MAPS OF CONSTANT RANK

Anthony Phillips

S.U.N.Y., Stony Brook

A smooth map $f : M \to W$ between two smooth manifolds is said to have constant rank k if at each $x \in M$ the differential $df|_x$ is a linear map of rank k. Associating to each such map its differential $df : TM \to TW$ gives a map $d : \mathrm{Hom}_k(M,W) \to \mathrm{Lin}_k(TM,TW)$ which is continuous w.r.t. the C^1 and compact-open topologies.

Theorem. If M is an open manifold, and has the homotopy type of a complex of dimension $<k$, then d is a weak homotopy equivalence.

This generalizes (for open manifolds) the Smale-Hirsch classification of immersions.

AN EXERCISE IN HOMOLOGY THEORY AND AN APPLICATION TO THE FOUR COLOUR CONJECTURE AND OTHER COLOURING PROBLEMS

S. Thomeier

Memorial University, St. John's

An admissible four-colouring of a normal map on the two-dimensional sphere can be regarded as a 2-chain on S^2 with coefficients in the cyclic group Z_4 of order 4 or in the Klein group $Z_2 \oplus Z_2$.

By adopting this viewpoint one can give a trivial proof of the well-known Wolynski-Tait Theorem (cf. S. Thomeier, Lectures on Homology Theory, Aarhus Matematisk Institut Lecture Notes, 1967). In a similar manner one can translate the Four Colour Conjecture into various equivalent statements about colourings of the edges or vertices of such maps with three or two colours.

Analogous such translations are also possible for the case of maps on surfaces of higher genus; these are, of course, less interesting since the colouring problem for such maps has already been completely solved. It is, however, interesting to apply the same idea to colouring problems in higher-dimensional simplicial (or cellular) complexes.

THE METASTABLE HOMOTOPY OF CLASSICAL GROUPS AND WHITEHEAD PRODUCTS

S. Thomeier

Memorial University, St. John's

It is clear that questions concerning the homotopy of spheres, of the classical groups, and of other important spaces such as Stiefel and Grassmann manifolds are all interrelated via the various fibrations connecting these spaces. It is equally clear that such questions can also be expressed in terms of Whitehead products on spheres. In particular, any information one has on the order, divisibility, or desuspensibility of such products can readily be transferred to obtain corresponding information on the homotopy of the above mentioned spaces. The following theorem is only a particularly simple example.

Let k be 1, 2, or 4 (for the real, complex, or quaternionic case) and let $G(n)$ stand for the orthogonal, unitary, or symplectic groups $O(n)$, $U(n)$, or $Sp(n)$, respectively. Denote by $i: G(n-1) \to G(n)$ the usual inclusion map and let $\iota_m \in \pi_m(S^m)$ be the identity class element. Considering the fibration $G(n) \to S^{kn-1}$, a standard obstruction argument gives the following

<u>Theorem</u>: If $i_*: \pi_{kn+r-2}(G(n-1)) \to \pi_{kn+r-2}(G(n))$ is a monomorphism, then the Whitehead product $[\alpha, \iota_{kn-1}]$ is zero for all $\alpha \in \pi_{kn+r-1}(S^{kn-1})$.

Combining this and similar results with relations connecting the behaviour of different Whitehead products (cf., for instance, S. Thomeier, Proceed. of the 13th Biennial Sem. Canad. Math. Congress, vol. <u>2</u> (1971), 144-155), one gets interesting consequences about the homotopy groups of the above mentioned spaces in the metastable range. These are, of course, also related to properties of the unstable J-homomorphism.

A GENERALIZATION OF NOVIKOV'S THEOREM TO FOLIATIONS WITH ISOLATED GENERIC SINGULARITIES

Edouard Wagneur

Université de Montréal

In this paper, we want to give sufficient conditions for the generalization of Novikov's theorem [3] to foliations of S^3 with isolated generic singular points.

I. INTRODUCTION

Let M be a C^∞ connected, 3-dimensional manifold and ω a C^∞ completely integrable 1-form on M (i.e. $\omega \wedge d\omega = 0$). Generically, the singularities of ω are isolated points and 1-dimensional submanifolds of M (see [1] for more details). If we assume that ω remains transversal to the 0-section of the cotangent bundle $T^*M \to M$ the singularities of ω consist of isolated points only and the rank of the Jacobian matrix of ω at such a point is maximal.

The integral manifolds of ω determine a codimension one foliation of M with isolated singular points corresponding to the singularities of ω. Such foliations have been studied in [6]. For simplicity, we will say foliation for foliation with isolated generic singularities.

Let us recall the situation for the possible singular points:

i) <u>the spherical point</u>

In a small nbhd of such a point, the leaves of the foliation are diffeomorphic to the 2-sphere S^2 [4].

ii) <u>the conical point</u>

In an appropriate nbhd U of such a point, there is a "cone" whose vertex is the singular point of the foliation. The cone separates U into three connected components (see the figure). The two "parabolic" components are foliated by open discs.

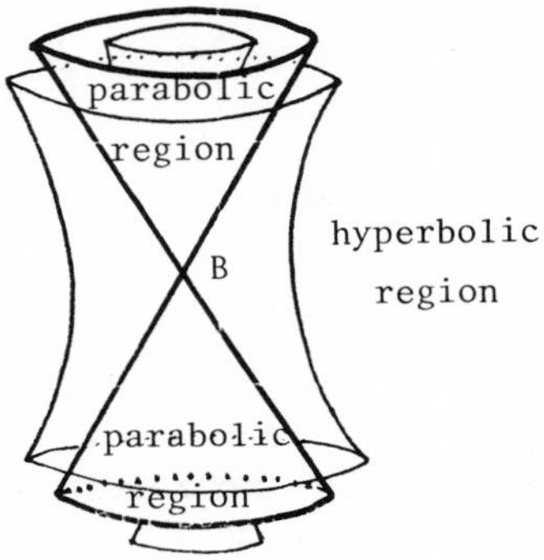

The third component , called "hyperbolic" may be foliated as indicated in a), b) or c) below.

a) All the leaves are diffeomorphic to the cylinder $S^1 \times]0, 1[$.

b) All the leaves are diffeomorphic to the plane $R \times]0,1[$, where the R component spirals toward the cone.

c) The leaves consist of cylinders (like in case a)) on the one hand and of planes $R \times]0,1[$ (like in case b)) where the R component spirals towards two cylinders, on the other hand.

These three cases for a conical singular point will be refered to as type a), b) or c) respectively.

Also, the vertex of the cone (the conical singular point) is considered to belong to both branches of the cone. A singular leaf F of a foliation $\mathcal{F}$ may then be defined as a leaf F containing (conical) singular points $B_1, B_2, \ldots$ such that $F - \bigcup B_i$ is a (connected) leaf of the regular foliation induced by $\mathcal{F}$ on $M - S$, where S is the set of all singular points of $\mathcal{F}$.

For compact M, we have $\chi(M) = 0$ (the Euler characteristics of M). Using a vector field transversal to the foliation $\mathcal{F}$ (or transversal to the 1-form ω defining $\mathcal{F}$) we can define the index of a singular point of $\mathcal{F}$ to conclude the total number of singular points must be even. We write s and c respectively for the number of spherical and conical singular points of F.

In the next paragraph, we show how singularities may be reduced two by two. Assuming this, we can define what is meant by a Reeb component of a (singular) foliation of M.

Definition. A Reeb component of a foliation F of M is a submanifold R of M such that

i) R is diffeomorphic to the solid torus $D^2 \times S^1$

ii) $\partial R = R - \mathrm{Int}(R)$ is a leaf of F

iii) $F|R$ can be reduced to the classical Reeb foliation of the solid torus.

The main result of this paper can now be stated.

Theorem. If F is a foliation of S^3 with $s = c$ then, generically, F admist a Reeb component.

Remarks: 1. The classical Novikov's theorem states that our theorem is true when $s = c = 0$.

2. In [5], H. Rosenberg gives an example showing that there exist foliations of S^3 with $s < c$ and no compact leaf.

3. In [6], we prove that when $s > c$ then $(s = c + 2$ and) generically all the leaves are homeomorphic to S^2.

The next two paragraphs summarize results of [6]. Some details needed for the proof of our theorem are given.

II. REDUCTION OF SINGULARITIES

Let F be a foliation of M (compact) with $s \geq 1$, $c \geq 1$. For every spherical point A, define $\Omega(A)$ to be the union of all 3-discs D_i with $A \in D_i$ $(i \in I)$ such that $\partial D_i \simeq S^2$ is a leaf of F and $F|D_i$ is diffeomorphic to the foliation of the unit disc by 2-spheres with a spherical singular point at the origin $(i \in I)$. It is easy to see that $\Omega(A)$ is diffeomorphic to the 3-ball and that (if $c \geq 1$) $\overline{\Omega}(A)$ contains at least one conical singularity.

The situation of a conical singular point $B \in \overline{\Omega}(A)$ falls into one of the following three cases. Let U be an appropriate nbhd of B and denote by $P_i(U)$, i = 1,2 (resp. $H(U)$) the two parabolic regions (resp. the hyperbolic region) of B in U:

Case 1. $\Omega(A) \cap U = P_1(U)$. In this case we can prove that B belongs to two distinct singular leaves. Up to other singular (conical) points lying on $\partial\overline{\Omega}(A) = \overline{\Omega}(A) - \Omega(A)$ this situation can be represented by the following rotation figure.

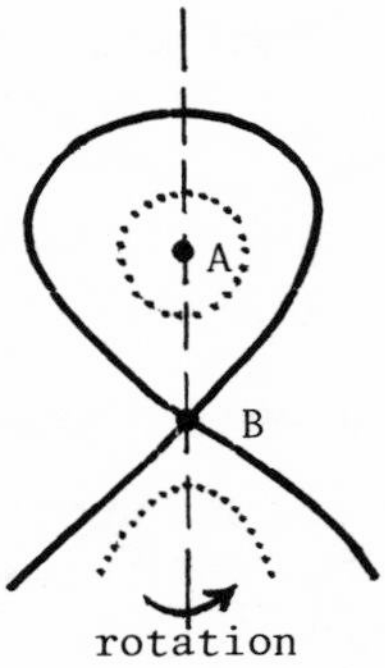

Case 2. $\Omega(A) \cap U = H(U)$. Again in this case, one can show that B belongs to two distinct singular leaves. Up to other singularities of $\partial\overline{\Omega}(A)$ this situation is similar to the one given previously (rotation around the axis).

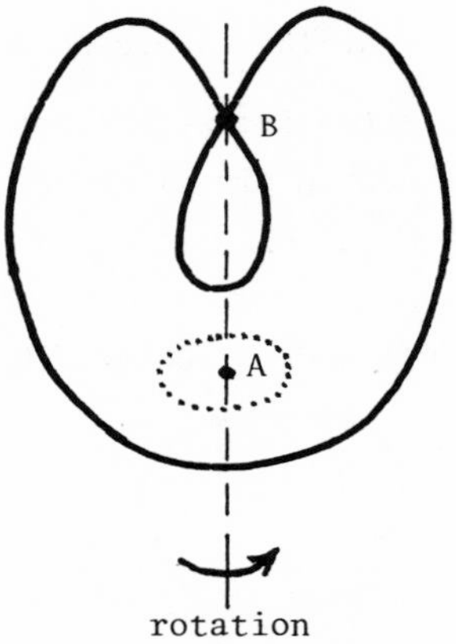

Case 3. $\Omega(A) \cap U = P_1(U) \cup P_2(U)$. The conical singularity B can be proved to belong to one leaf only. Up to other singular points of $\partial\overline{\Omega}(A)$, $\overline{\Omega}(A)$ looks like the pinched torus on the following page.

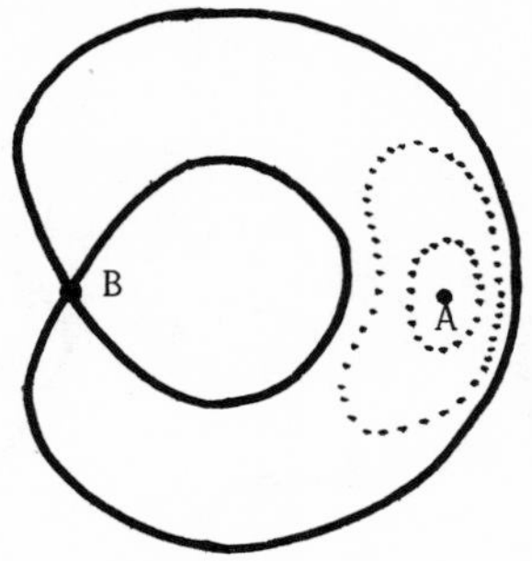

In [6] we show how, on each case, the singular points A and B of F can be eliminated, generating a new foliation F' of M with $s' = s - 1$, $c' = c - 1$ such that $F'|M - K'$ is diffeomorphic to $F|M - K$, where K and K' are two compacts of M.

Here is a geometric description of the reductions occuring in the first two cases.

Case 1. Let U be an appropriate nbhd of B and $\phi : D^2 \times [0,1] \to U$ an imbedding such that

i) $\phi(D^2 \times [0,1])$ is a (compact) nbhd of B

ii) for small ε, $\phi \mid \partial D^2 \times [0,\varepsilon[$ and $\phi \mid \partial D^2 \times]1 - \varepsilon,1]$ are imbeddings into the two singular leaves $F_0 = \partial\overline{\Omega}(A)$ and F_1 containing B.

iii) The induced foliation on $D^2 \times \{i\}$ $(i = 0,1)$ is the foliation of D^2 with one singular point : a center at the origin.

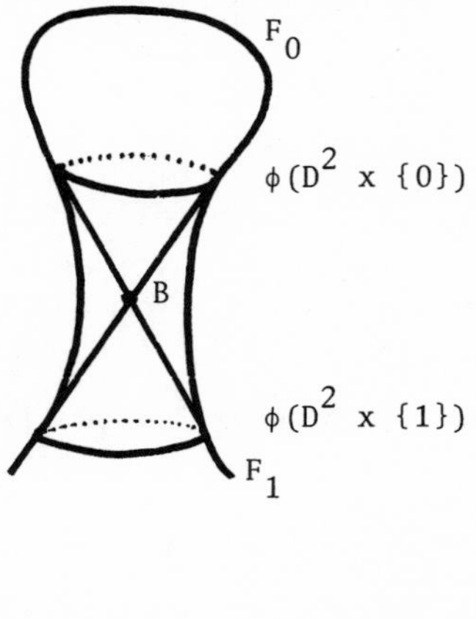

Take $L \subset \mathcal{P}_2(U)$ to be the compact with boundary $\partial L = C \cup \phi(D^2 \times \{1\})$ where $C = F_1 \cap \phi(D^2 \times [0,1])$. Let $K = \overline{\Omega}(A) \cup L$, $K' = \overline{\Omega}(A) \cup \phi(D^2 \times [0,1])$, and N be a small nbhd of $\phi(D^2 \times \{1\})$ in M such that $\phi(D^2 \times \{1 - \frac{\varepsilon}{2}\} \cap N = \emptyset$.

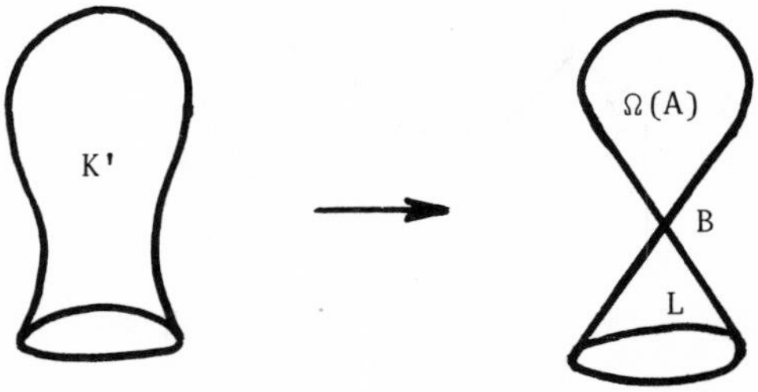

Clearly there is a diffeomorphism

$$\Psi_1 : M - K' \to M - K$$

which reduces to identity on $N \cap (M - K')$
and a diffeomorphism

$$\Psi_2 : \mathrm{Int}(K') \to \mathrm{Int}(L)$$

which reduces to identity on $N \cap \mathrm{Int}(K')$.
The new foliation $\mathcal{F}'$ on M is then defined by

$$\mathcal{F} \mid (D^2 \times \{1\})$$

$$\Psi_1^{-1}(\mathcal{F} \mid M - K) \quad \text{on} \quad M - K'$$

$$\Psi_2^{-1}(\mathcal{F} \mid \mathrm{Int}(L)) \quad \text{on} \quad \mathrm{Int}(K')$$

and by adding a new leaf

$$F' = \Psi_1^{-1}(F_2 - C) \cup \phi(\partial D^2 \times \{1\}) \cup (\partial K' - \phi(D^2 \times \{1\})).$$

Case 2. The imbedding ϕ of the cylinder $D^2 \times [0,1]$ into a nbhd U of B is chosen in such a way that

i) $\phi \mid T \times \{i\}$ is an imbedding of a tubular nbhd T of ∂D^2 in D^2 into the singular leaf F_i containing B, $i = 0,1$.

ii) for small ε, $\phi(\partial D^2 \times]\varepsilon, 1 - \varepsilon[)$ belongs to a leaf $F \subset \Omega(A)$.

Take $K = \overline{\Omega}(A)$, $K' = \overline{\Omega}(A) \cup \phi(D^2 \times [0,1])$.

It is easy to show that $\mathrm{Int}(K') \simeq S^2 \times]0,1[$ and that there exists a diffeomorphism

$$\Psi : M - K' \to M - K.$$

$\mathcal{F}'$ is defined by

$$\Psi^{-1}(\mathcal{F} \mid M - K) \text{ on } M - K'$$

the trivial foliation $S^2 \times \{t\}$ on K'

taking the two connected components F_0' and F_1' of $\partial K' = K' - \mathrm{Int}(K')$ as leaves.

Remarks: 1. In case 2, if B is the only singular point of $\partial\overline{\Omega}(A)$ then F_0' and F_1' can be shown to be diffeomorphic to S^2. As a consequence of Reeb's stability theorem [4] neighbour leaves are also diffeomorphic to S^2. Hence, for some nbhd U of B, every leaf which meets $\mathcal{P}(U)$ is also diffeomorphic to S^2, $i = 1,2$.

2. In each case, the reduced foliation $\mathcal{F}'$ is independant of the choice of the compact K.

III. GENERIC SITUATION OF SINGULAR POINTS

Let $\sigma(M)$ be the space of completely integrable 1-forms on M which are transversal to the 0-section of the cotangent bundle $T^*M \to M$ with C^1-topology. According to the classical terminology, a subset S of $\sigma(M)$ is residual if it contains a countable inter-section of dense open subsets of $\sigma(M)$. A generical property for 1-forms is a property which defines a residual subset of $\sigma(M)$. Also for foliations, genericity will always refer to the corresponding situation for 1-forms.

In [6] we prove the following

Theorem 1. Let $\mathcal{F}$ be a foliation of M. Generically, every (conical) singular point belongs to two distinct leaves and every singular leaf contains exactly one singularity (necessarily conical).

Corollary. Generically, for every spherical singular point A, there is only one conical point $B \in \partial\overline{\Omega}(A)$ and this point is in the situation of Case 1 or 2 of §II.

Remark. The proof of theorem 1 uses, in a neighbourhood U of a singular point B, a small C^1-perturbation of the foliation in U. Such a perturbation removes the point B from a given leaf to put it on another leaf. Geometrically this perturbation can be represented as follows (see [6] for more details).

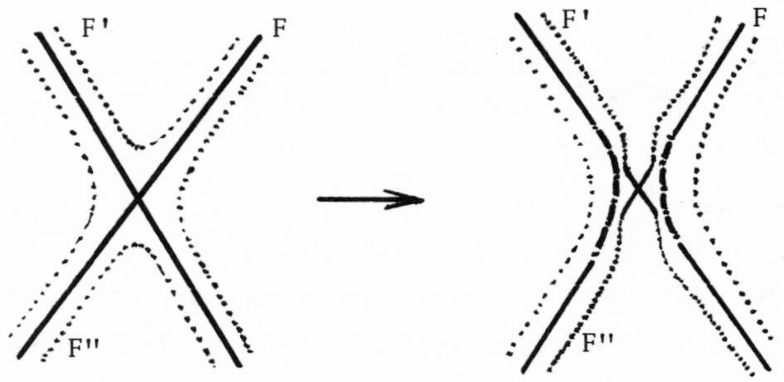

The smoothing of F makes the two leaves F' and F'' singular at B.

IV. THE THEOREM OF NOVIKOV

In this paragraph, we always deal with the generical case, in particular every foliation satisfies the conclusion of Theorem 1.

Lemma. Let $\mathcal{F}$ be a foliation of M with two singular points A and B (respectively spherical and conical) such that $B \in \overline{\Omega}(A)$. If $\mathcal{F}'$ is the foliation of M given by the reduction of singularities A and B by the method of II then generically every Reeb component of $\mathcal{F}'$ corresponds to a Reeb component of $\mathcal{F}$.

Proof. Let R' be a Reeb component of $\mathcal{F}'$ and K' be the compact used in the reduction of singularities A and B.

If $R' \cap K' = \emptyset$ then ∂R is a leaf of $\mathcal{F}$ and either R' is a Reeb component of $\mathcal{F}$ or $\mathcal{F}'|R'$ is the reduced foliation of $\mathcal{F}|R$, where R is a Reeb component of $\mathcal{F}$ with $\overline{\Omega}(A) \subset R$.

If $R' \cap K' \neq \emptyset$, by the independance of the choice of K' we may suppose that $\partial R'$ comes from a leaf $F \in \mathcal{F}$ which contains B. Also, by the remark 1 in §II, the reduction must have been that of case 1 and $(\partial R' - K') \subset F$. Since $(\partial R' - K') \simeq F - K$ (with $\partial R' \cap K'$ and $F \cap K$ both homeomorphic to D^2) F is homeomorphic to $\partial R'$. The "inclusion" $\partial R' - K' \subset F$ implies that F is the boundary of a solid torus in M. Now generically, using the perturbation of §III, the singular point B can be removed from F. Hence generically F is a regular leaf of $\mathcal{F}$ bounding a Reeb component R of $\mathcal{F}$.

Theorem 2. Let $\mathcal{F}$ be a foliation of S^3 with $s = c$. Then generically $\mathcal{F}$ admits a Reeb component R with $\partial R \simeq S^1 \times S^1$ (diffeom.).

Proof. By induction on c.

For $c = 0$ this is the classical result of Novikov [3]. Suppose the theorem is true for every $c' < c$ and let A and B be two singular points of $\mathcal{F}$ with A spherical and $B \in \overline{\Omega}(A)$. If $\mathcal{F}'$ is the foliation on S^3 given by the reduction of singularities A and B then by induction hypothesis $\mathcal{F}'$ admits a Reeb component R' with boundary $\partial R' \underset{\circ}{\simeq} S^1 \times S^1$. By the lemma above $\mathcal{F}$ admits a Reeb component R.

Remarks: 1. The reason why we deal with the generical case is that when reducing the singularities in case 3 of the reduction lemma we create a Reeb component. Also this situation of case 3 is not generic for the conical singular point B belongs to only one leaf.

2. It is then natural to ask whether the theorem still holds for every foliation of S^3 with $s = c$. As shown by the following example this is generally not the case.

Example. Let $\mathcal{F}$ be a regular foliation of S^3 with a knotted Reeb component R. According to [2], $S^3 - R$ may be foliated by planes with a handle.

If instead of R we take a pinched knotted solid torus T then since $S^3 - T \simeq S^3 - R$ there is a singular foliation of S^3 with $s = c = 1$ and no Reeb component. This example can easily be generalized to any $s = c > 1$.

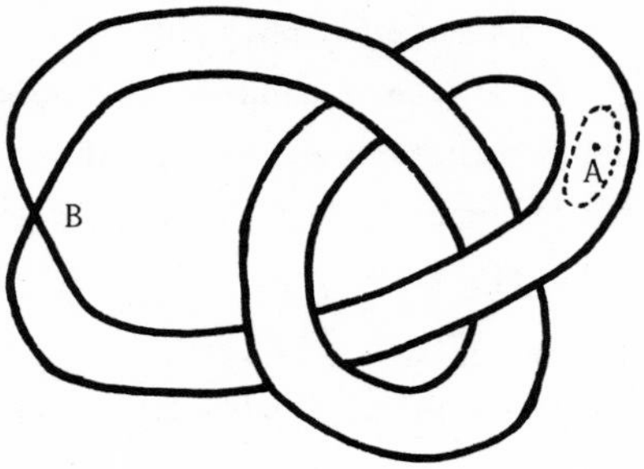

REFERENCES

[1] I. Kupka, to be published.

[2] B. Lawson, Codimension one foliations of spheres, Ann. of Math., 94 (1971), 494-503.

[3] S.P. Novikov, Topology of foliations, Trudy Mosk. Math. ob. 14 (1965), 367-397.

[4] G. Reeb, Sur certaines propriétés topologiques des variétés feuilletées, A.S.I. 1183, Paris, 1952.

[5] H. Rosenberg, La théorie qualitative des feuilletages. Sem. Math. Sup. 1972. Université de Montréal.

[6] E. Wagneur, thèse, Université de Montréal, 1973.

HOMÖOMORPHIE EINIGER TYCHONOFFPRODUKTE

Ernst Witt

Universität Hamburg

$\mathcal{B}$ bezeichne die Klasse der topologischen Räume, die kompakt, total unzusammenhängend und metrisch sind und mindestens zwei Punkte haben. $\mathcal{B}$ enthält nur einen perfekten Raum

$$C = \{0,1\}^{\mathbf{N}} \simeq \{\sum_1^\infty \varepsilon_i \, 3^{-i} \in \mathbf{R} \mid \varepsilon_i = 0,1 \} ,$$

das sogenannte Cantorsche Diskontinuum [Willard, General Topology (1970), 210-219]. Insbesondere ist jedes abzählbar unendliche Produkt $\Pi B_i \simeq C$, ($B_i \in \mathcal{B}$), und man hat Injektionen $B_i \to C$. Aut C ist transitiv (man interpretiere C als Gruppe), aber sogar m-fach transitiv, denn Aut C vertauscht m Punkte von C ($m<\infty$) symmetrisch, wie man durch Vertauschung von disjunkten offen-abgeschlossenen Umgebungen ($\simeq C$) dieser Punkte erkennt.

Jede unendliche hausdorffsche kompakte Gruppe G mit einer abzählbaren Kette offener Untergruppen U_i als Umgebungsbasis der 1, ($U_1 = G$), ist homöomorph C, denn U_i / U_{i+1} ist endlich,und topologisch ist $G \simeq \Pi\, U_i / U_{i+1} \simeq C$.

K_p sei ein p-adischer Körper mit endlichem Restklassenkörper, o der Ring ganzer Zahlen, p das Primideal, $\overline{K}_p = p \cup \frac{1}{o} = K_p \cup \infty$.

Beispiele für $G \simeq C$: (1) Die additiven Gruppen von p und o mit der Basis p^i, (2) die Einheitengruppe U mit der Basis $1+p^i$, (3) eine unendliche Galoisgruppe von L/K, wenn K in L nur abzählbar viele endliche Erweiterungen besitzt (das trifft zu, wenn K absolut algebraisch ist oder für $K = K_p$). Ferner ist topologisch

$$\overline{K}_p \simeq C \cup C \simeq C , \qquad K_p \simeq \dot{C} ,$$

dabei entstehe $\dot{C}$ aus C durch Punktieren, d.h. durch Entfernen eines beliebigen Punktes.

Die verschiedenen p-adischen Körper K_p unterscheiden sich also topologisch überhaupt nicht. Die Frage nach der Möglichkeit homöomorpher K_p, die mir kürzlich S. Thomeier gestellt hatte, war Anlass zu dieser Note.

Bevor ich das inhaltsreiche Buch von Willard zu Rate zog, fand ich einen direkten Beweis von $K_p \simeq \dot{C}$ durch explizite Angabe der folgenden Homöomorphie

$$f : B = \Pi B_i = \Pi\{0,1,\ldots,n_i\} \to C \ , \quad (n_i > 0,\ i \in \mathbf{N}) \ ,$$
$$f(b) = \Pi\ (1^{b_i}\ 0^{\delta(b_i<n_i)}) \ ,$$

hierbei sei $\delta(\xi) = 1$, wenn ξ zutrifft, sonst $= 0$. Zum Beispiel bedeutet $\delta(i=k)$ das Kroneckersymbol.

Beweis der Homöomorphie: $c_1 \ldots c_n$ hängt nur von $b_1 \ldots b_n$ ab, daher ist f gleichmässig stetig. $f^{-1}(c)$ setzt sich entsprechend $B = \{\Pi\ (1^{q_k}\ 0)\ \Pi\ 1\}$ zusammen: $q_1 = n_1 + \ldots + n_{s-1} + r$, $(0 \le r < n_s,\ s \ge 1)$, ergibt $b_i = n_i$ für $i < s$, $b_s = r$. Daher ist f bijektiv. B ist kompakt und C hausdorffsch. Bekanntlich folgt jetzt $f : B \simeq C$.

Am Rande erwähnt sei auch folgende Homöomorphie

$$\Pi\ A_i = \Pi\ \{i,\ldots,\infty\} \to C \ , \qquad a \mapsto c \ :$$

Man setze $n = 0$, durchlaufe die Paare (i,k) mit $k \le i$ lexikographisch und setze nach jedem Schritt $n .= n + \delta(a_k \ge i)$, insbesondere $n .= n+1$, falls $k = i$, und $c_n\ \delta(a_k \ge i) = \delta(a_k = i)$.

AUTHOR INDEX

Numbers refer to pages on which a reference is made to an author or a work of an author. *Italic* numbers indicate the first pages of the articles in this volume.